Predictive Maintenance:

Transforming Industrial Operations for Tomorrow

Dr. Gregory Thompson

Ronald Wagner

Robert Ufford

ISBN: 979-8-9899555-2-7

FOREWORD

Why did we write this book?

Two decades ago, a computer scientist with a sensor background, an engineer with a communications background, and a mathematician with a maintenance background began collaborating on projects to extend the life of complex equipment.

- We developed an integrated sensor and radio frequency device to remotely monitor conditions inside high value equipment containers preventing moisture damage in shipping and storage.
- We competed in a comparison of tracking solutions to monitor and report conditions of containers shipped by truck, rail, and ship from Indiana, to San Francisco, to the island Guam. We achieved 98 % coverage, the highest of the three competing solutions.
- We improved maintenance on metropolitan transit buses, using existing CAN bus data. We discovered higher HVAC failures on buses that more frequently opened and closed their doors.
- We pioneered sensors to position snowplow blades at the optimum height for conditions on the roads.
- We configured and integrated a work-in-progress inventory visibility system for a major aerospace manufacturing plant.
- We developed and patented a device to convert analog data to digital to provide battlefield commanders with a software information application that could predict performance probability among the five capabilities: "Move, Shoot, Communicate, Navigate, Survive".
- We developed and prototyped a GPS based tracking system that communicated via cellular the precise location of repair vehicles to reduce "windshield time"- the unnecessary driving to/from repairs without the correct tools.

Along the way, the engineer who writes everything down, compiled a compendium of tools, techniques, applications, and benefits of what we call "Predictive Maintenance". This book is our diary of what we used and will use in the pursuit of excellence in improved sustainment of systems and processes in maintenance and performance. We hope you find it as useful as we have.

Table of Contents

List of Figures

Part I: Introduction to Predictive Maintenance

Chapter 1: Introduction to Maintenance Strategies

Maintenance is an essential component of every industrial operation. Ensuring the smooth and efficient functioning of machinery, systems, and tools lies at the heart of successful and sustainable business performance.

Businesses can adopt several maintenance strategies, each with its unique benefits and limitations. Understanding these strategies – reactive, preventive, conditioned-based, and predictive maintenance – is crucial for making informed decisions about which approach to apply and when.

Professionals and academics have approached the maintenance challenge in several ways. Quite a few terms have surfaced for similar types of maintenance. For example, the term reactive maintenance is often called corrective maintenance. There can be offshoots of maintenance types.

An example of this is reliability-centered maintenance. It is a specialized methodology that determines the periodic intervals for conducting preventive maintenance. Thus, it behooves us to define the terms used in this volume to ensure that the focus remains on predictive maintenance.

Reactive Maintenance (RM), also known as breakdown or run-to-failure maintenance, is a strategy where maintenance work is performed after a piece of equipment fails or breaks down.

While this approach often results in higher repair costs and downtime because the problem isn't addressed until after the failure, it may be a suitable strategy for non-critical systems or equipment that's cheap or easy to replace. See Figure 1.

Preventive Maintenance (PM), on the other hand, involves routine inspections, servicing, and repairs to prevent potential failures before they occur. This approach aims to reduce the likelihood of unexpected breakdowns by carrying out maintenance activities on a pre-determined schedule, regardless of the equipment's condition. While preventive maintenance can reduce the risk of sudden failures, it can also result in unnecessary maintenance of equipment in good condition, leading to higher operational costs.

Predictive Maintenance (PdM) is a strategy that leverages data from equipment sensors and advanced analytics techniques to predict when a machine will fail or require service. The goal is to perform maintenance activities just in time to prevent the expected failure, minimizing downtime and costs. This method requires significant up-front investment in technology and skills, but it offers the potential for substantial cost savings and efficiency improvements in the long run.

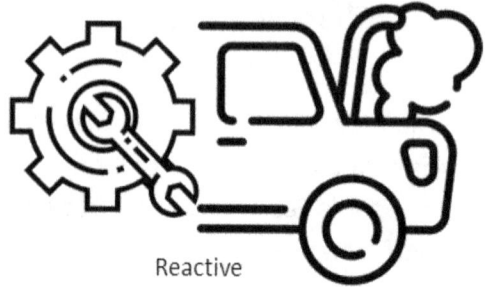

Reactive Maintenance

Preventive Maintenance

Predictive Maintenance

Figure 1: Maintenance Methods

Condition Based Maintenance (CBM) involves monitoring equipment performance and asset management with visual inspections, scheduled tests, and sensor devices to determine the most cost-efficient time to perform maintenance.

Conditioned Based Maintenance + (CBM+) is the application and integration of appropriate processes, technologies, and knowledge-based capabilities to improve the reliability and maintenance effectiveness of DoD systems and components. At its core, CBM+ is maintenance performed based on evidence of need.

Predictive & Prognostic Maintenance (PPMx) is applying and integrating appropriate processes, technologies, and knowledge-based capabilities to use authoritative and emerging data to achieve foresight in combat system health management and health management response.

The evolution of maintenance strategies reflects the progress in technological and analytical capabilities. Initially, businesses relied solely on reactive maintenance, fixing things only when they broke. The advent of scheduled maintenance procedures heralded the era of preventive maintenance. However, the real game-changer was the arrival of predictive maintenance, fueled by advancements in technologies such as the Internet of Things (IoT), machine learning (ML), and artificial intelligence (AI).

AI Predictive maintenance (AIPdM) is a technique that uses data and advanced analytics, including artificial intelligence (AI), to predict when equipment or machinery is likely to fail. By proactively identifying potential failures, organizations can schedule maintenance activities in advance, reducing downtime, minimizing costs, and optimizing resource allocation. Here's a general overview of how AI can be used for predictive maintenance:

1. **Data Collection:** The first step is to collect relevant data from the equipment or machinery that requires maintenance. This

includes sensor readings, performance metrics, operational logs, maintenance records, and other relevant data sources. The data should cover a sufficient time and include normal and failure conditions.

2. **Data Preparation:** Once the data is collected, it must be preprocessed and prepared for analysis. This involves cleaning the data, removing outliers, handling missing values, and converting it into a suitable format for analysis.
See Figure 2.

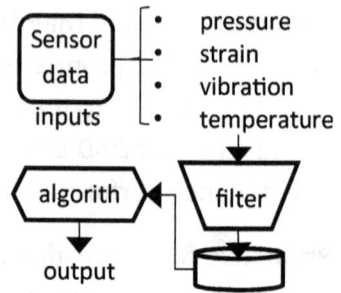

Figure 2: PdM Model Development

3. **Feature Extraction:** In this step, you extract meaningful features from the collected data. This can involve transforming the raw data into relevant features that capture the behavior and performance of the equipment. Feature extraction techniques can include statistical analysis, time-series analysis, Fourier transforms, or other domain-specific methods.

4. **Model Development:** After extracting the features, you must build a predictive model using AI techniques. Several AI algorithms can be used, such as machine learning algorithms (e.g., regression, decision trees, random forests, support vector machines) or deep learning algorithms (e.g., recurrent neural networks, convolutional neural networks). The algorithm's choice depends on the problem's complexity and the available data.

5. **Model Training:** The predictive model needs to be trained using historical data, where the input features are used to predict the equipment's failure or maintenance needs. The

model is adjusted and optimized based on the training data, ensuring it captures the patterns and relationships between the input features and failure events.

6. **Model Evaluation:** Once the model is trained, it must be evaluated to assess its performance. This is typically done using test data not used during the training phase. Evaluation metrics such as accuracy, precision, recall, and F1-score can be used to measure the model's performance. If the model's performance is not satisfactory, further iterations of training and evaluation may be required.

7. **Deployment and Monitoring:** After the model is evaluated and deemed adequate, it can be deployed in a production environment. Real-time data from the equipment or machinery is continuously fed into the model, which makes predictions about future failures or maintenance requirements. These predictions can be used to generate alerts, trigger maintenance activities, or provide recommendations to operators.

8. **Continuous Improvement:** Predictive maintenance is an ongoing process. As new data is collected and the model operates in the real world, it's essential to monitor its performance and make necessary adjustments continuously. This may involve retraining the model periodically with new data or updating it to incorporate domain-specific knowledge or changes in the equipment or operational conditions.

By leveraging AI for predictive maintenance, organizations can optimize maintenance schedules, reduce unplanned downtime, extend equipment lifespan, and improve operational efficiency. Today, forward-looking companies are leveraging these technologies to shift towards predictive maintenance, heralding a new era in industrial operations management.

The PdM goal is to establish an environment where reactive, unscheduled maintenance can be replaced by predictive maintenance (PdM) executed at the most reasonable time and location with the right people, correct parts, and best tools. A quick overview and peek ahead are provided in Figure 3.

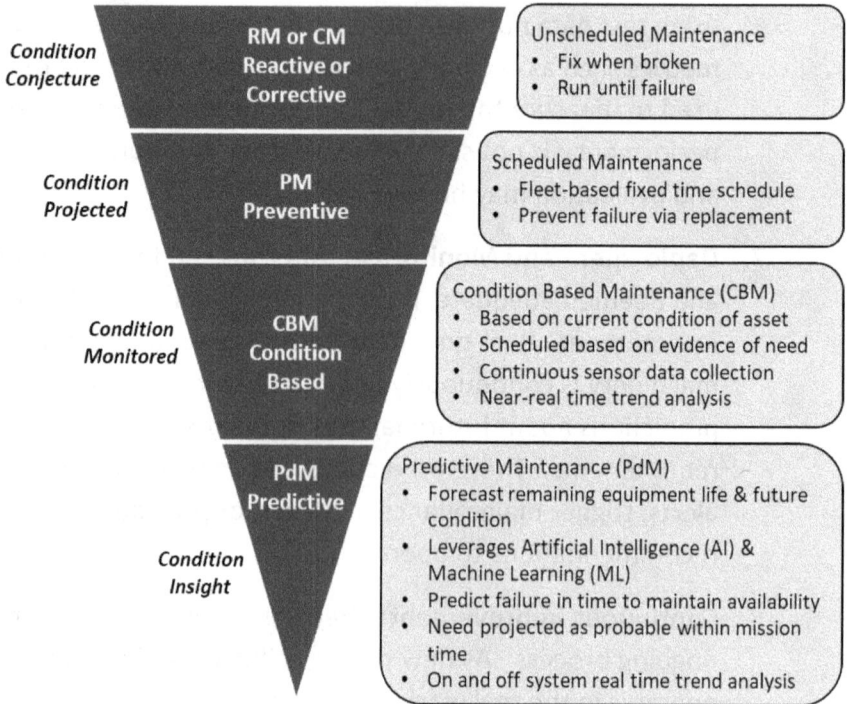

Condition Conjecture	RM or CM Reactive or Corrective	Unscheduled Maintenance • Fix when broken • Run until failure
Condition Projected	PM Preventive	Scheduled Maintenance • Fleet-based fixed time schedule • Prevent failure via replacement
Condition Monitored	CBM Condition Based	Condition Based Maintenance (CBM) • Based on current condition of asset • Scheduled based on evidence of need • Continuous sensor data collection • Near-real time trend analysis
Condition Insight	PdM Predictive	Predictive Maintenance (PdM) • Forecast remaining equipment life & future condition • Leverages Artificial Intelligence (AI) & Machine Learning (ML) • Predict failure in time to maintain availability • Need projected as probable within mission time • On and off system real time trend analysis

Figure 3: Pathway to Predictive Maintenance

Chapter 2: The Importance of Predictive Maintenance

Predictive maintenance has emerged as a crucial strategy for modern industries due to its substantial impact on reducing the costs and risks associated with unplanned downtime. Unplanned downtime refers to unexpected incidents that halt production, such as machine failures, network disruptions, or supply chain interruptions. These incidents can be immensely disruptive, leading to significant production loss, missed delivery timelines, decreased customer satisfaction, and increased emergency repairs or replacement costs. See Figure 4.

Figure 4: Importance of Predictive Maintenance

One study by Aberdeen Research estimated that unplanned downtime can cost a company as much as $260,000 per hour. Beyond this immediate financial impact, long-term costs are associated with a damaged reputation, lost customers, and lower market share. Reactive maintenance, while having its place in a comprehensive maintenance strategy, can lead to more of these unplanned outages because it only addresses equipment issues after a failure has occurred.

This is where predictive maintenance steps in, offering benefits that directly counter the costs and risks of unplanned downtime. By analyzing data collected from various sources, predictive maintenance can identify potential issues and vulnerabilities in the equipment before they escalate into major problems. This approach allows organizations to schedule maintenance activities during non-productive times, thereby reducing disruptions to production. Moreover, predictive maintenance can optimize the use of resources by only focusing on equipment that needs attention instead of preventive maintenance, which can sometimes lead to unnecessary work on machines that are functioning correctly.

In addition to reducing downtime and associated costs, predictive maintenance can extend the life of machinery by catching minor issues before they become significant problems. This approach can also improve operational efficiency by allowing organizations to better plan their maintenance workforce, parts inventory, and equipment usage. Furthermore, insights derived from predictive maintenance can help organizations make more informed decisions about asset management, such as when to repair, replace, or retire machinery.

The adoption of predictive maintenance signifies a strategic shift from 'fix it when it breaks' to 'fix it before it breaks,' thus providing organizations with a competitive advantage in an increasingly challenging business environment. The importance of predictive maintenance will only grow as industries continue to digitalize and the cost of downtime becomes more critical. See Figure 5.

Predictive maintenance involves using data analysis techniques and machine learning algorithms to predict when equipment or machinery will likely fail, allowing for proactive maintenance actions. Here are some examples of predictive maintenance at work:

1. **Vibration analysis:** Sensors are placed on rotating equipment such as motors, pumps, and turbines to measure vibrations. By analyzing the vibration patterns over time, deviations from normal behavior can be detected, indicating potential faults or impending failures. Maintenance can then be scheduled before a breakdown occurs.

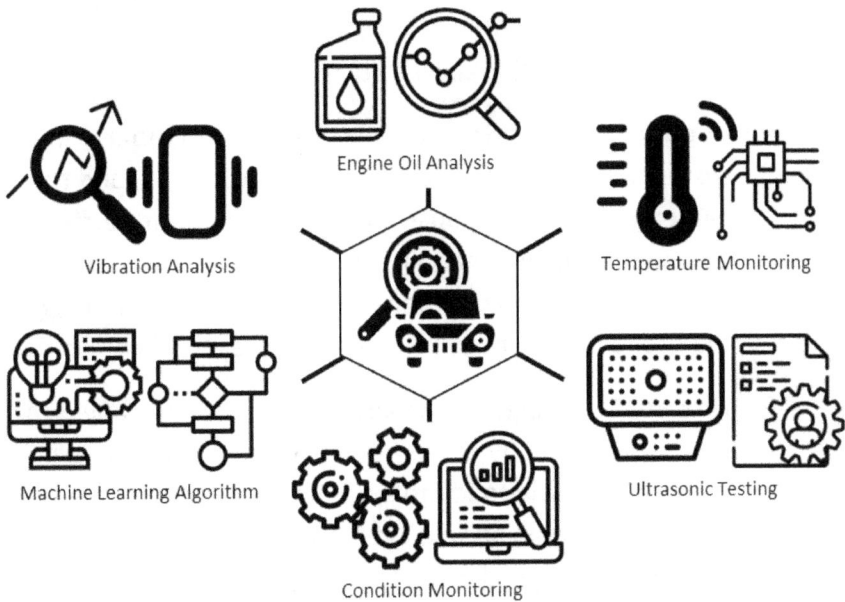

Figure 5: Predictive Maintenance for Vehicles

2. **Oil analysis:** Regularly analyzing the lubricating oil in machinery can provide valuable insights into the condition of components such as gears, bearings, and hydraulic systems. Changes in oil quality, contamination levels, or the presence of certain particles can indicate potential failures, allowing maintenance to be scheduled accordingly.

3. **Temperature monitoring:** Temperature sensors are installed in equipment to monitor the operating temperatures of critical components. Unusual temperature rises or fluctuations can indicate problems like friction, overheating, or inadequate cooling. Early detection helps prevent catastrophic failures and allows for targeted maintenance.

4. **Machine learning algorithms:** Advanced analytics and machine learning techniques can be applied to historical maintenance data, sensor readings, and other relevant parameters to identify patterns and correlations that can

predict future failures. These algorithms can learn from past failures and continually improve accuracy over time.

5. **Condition monitoring:** Various sensors and monitoring systems, such as pressure sensors, flow meters, and thermal cameras, can continuously monitor equipment performance and condition. By analyzing real-time data, deviations from normal operating conditions can be detected, enabling maintenance teams to intervene before a breakdown occurs.

6. **Ultrasonic testing:** Ultrasonic sensors detect changes in sound waves produced by equipment. Analyzing the sound patterns makes it possible to identify abnormalities such as leaks, mechanical wear, and defects. Maintenance actions can then be scheduled based on the severity of these abnormalities.

7. **Predictive analytics for vehicles:** In the automotive industry, predictive maintenance is used to monitor the health of vehicles and predict potential failures. Sensors and telematics devices collect data on parameters like engine performance, tire pressure, and battery condition. By analyzing this data, manufacturers and service providers can identify maintenance needs, schedule repairs, and improve vehicle reliability.

These examples demonstrate how predictive maintenance leverages data, sensors, and analytics to anticipate equipment failures, optimize maintenance schedules, reduce downtime, and enhance operational efficiency.

Several companies have successfully implemented predictive maintenance strategies to improve equipment availability and minimize downtime. See Figure 6. Here are a few real-life examples:

General Electric (GE): GE implemented predictive maintenance in their power generation division, using data analytics and machine learning algorithms to predict equipment failures. By analyzing sensor data from turbines and other critical components, GE can identify potential issues in advance and schedule maintenance activities proactively. This approach has increased equipment availability, reduced unplanned downtime, and improved operational efficiency.

Figure 6: Companies that have Successfully Utilized PdM

Boeing: Boeing utilizes predictive maintenance techniques in its aircraft operations. By collecting data from aircraft sensors, flight data recorders, and maintenance logs, they can monitor the health of various systems, such as engines, avionics, and landing gear. This data is analyzed to identify potential failures, allowing for timely maintenance actions and reducing aircraft downtime.

Rio Tinto: The mining company Rio Tinto uses predictive maintenance to optimize the availability of its mining equipment. They employ a range of sensors and data analytics tools to monitor the condition of critical machinery, such as trucks, excavators, and conveyors. By predicting equipment failures, Rio Tinto can schedule maintenance during planned downtime, ensuring that equipment is available when needed for mining operations.

Schindler: Schindler, a leading manufacturer of elevators and escalators, has implemented predictive maintenance to enhance equipment availability and reliability. They collect data from sensors installed in their elevator and escalator systems, analyzing them to detect anomalies, identify potential faults, and predict maintenance needs. This proactive approach enables Schindler to address issues before they result in breakdowns, improving equipment uptime and customer satisfaction.

United Airlines: United Airlines utilizes predictive maintenance to optimize the availability and performance of their aircraft. They collect vast amounts of data from various sources, including flight data recorders, onboard sensors, and maintenance logs. By applying advanced analytics techniques, they can identify patterns and trends that indicate potential failures or maintenance needs. This enables United Airlines to schedule maintenance tasks efficiently and reduce aircraft downtime, ensuring high availability for their fleet.

These real-life examples demonstrate how companies across different industries leverage predictive maintenance to improve equipment availability, reduce downtime, and enhance operational efficiency. By harnessing the power of data analytics and advanced technologies, these companies can proactively address maintenance needs, leading to increased reliability and cost savings.

Chapter 3: The Key Concepts Behind Predictive Maintenance

The innovative combination of traditional maintenance strategies and cutting-edge technologies such as data analysis and machine learning powers predictive maintenance. The synthesis of these fields provides a forward-looking, data-driven approach to equipment upkeep and efficiency optimization. Understanding these critical concepts behind predictive maintenance – data analysis, machine learning, predictive modeling, anomaly detection, and forecasting – can pave the way for more informed and effective implementation of this maintenance strategy.

Data Analysis is the process of inspecting, cleaning, transforming, and modeling raw data to discover useful information, draw conclusions, and support decision-making. In predictive maintenance, data analysis involves working with diverse types of data collected from various sources, such as sensor readings, historical maintenance records, and machine logs. This data is then analyzed to identify patterns, correlations, and trends that can help predict equipment failures. See Figure 7.

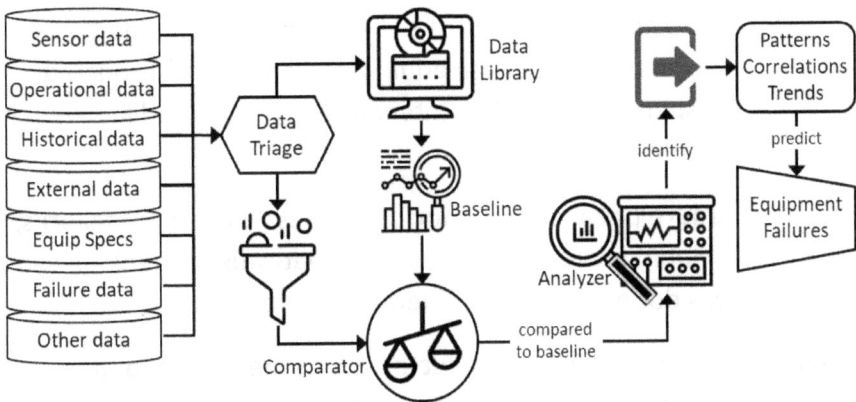

Figure 7: Data Analysis Process for PdM

The specific inputs required for data analysis in predictive maintenance vary greatly depending on the nature of the machinery or system under observation. However, typically, they can be broadly categorized as follows:

1. **Sensor Data:** These are real-time data gathered from various sensors installed on the equipment or machinery. Sensor data can include measurements of temperature, pressure,

humidity, vibration, acoustics, current, voltage, rotational speed, and other physical or electrical properties.

2. **Operational Data**: This could include information about the equipment's usage, such as its operating speed, load, duty cycles, and other operation-specific parameters.

3. **Maintenance History**: Data about past repairs, part replacements, and other maintenance activities. This data can provide valuable context and help identify trends or patterns indicating impending failures.

4. **External Factors**: Depending on the context, environmental factors like weather conditions or information about the supply chain (e.g., the quality of parts) may be relevant.

5. **Equipment Specifications**: Manufacturer-provided information about the equipment, such as its expected lifespan, suggested maintenance schedule, and known issues or vulnerabilities.

6. **Failure Data**: Information about past equipment failures, including when they occurred, what caused them, and what was done to fix the problem.

All these data are typically collected over time to form a time series data, which is crucial for predictive maintenance tasks. By examining these data inputs, patterns and correlations can be found and used to predict future equipment failures. Remember, these are general categories of inputs, and the specific data required will depend heavily on the type of equipment or machinery, the operating conditions, and the particular objectives of the predictive maintenance program.

Machine Learning is a subset of artificial intelligence that allows systems to automatically learn and improve from experience without being explicitly programmed. It relies on algorithms that can learn from and make predictions or decisions based on data. In predictive maintenance, machine learning algorithms can be trained on historical data to learn the complex patterns and relationships that lead to equipment failure, enabling these algorithms to predict future failures based on new data. See Figure 8 and the steps below.

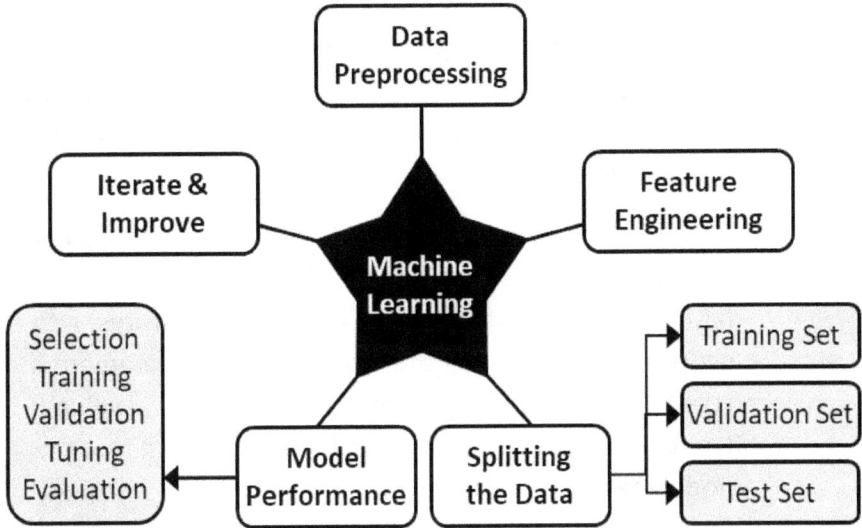

Figure 8: Machine Learning Elements & Features

Training machine learning algorithms on historical data involves a series of steps:

1. **Data Preprocessing**: This first step involves preparing the data for the machine learning algorithm. This may include cleaning the data (removing or imputing missing values, handling outliers), normalizing or standardizing the data (such as scaling the data to have a mean of 0 and standard deviation of 1), encoding categorical variables (turning category labels into numerical values), and balancing the dataset (in case of uneven distribution of classes).

2. **Feature Engineering**: This involves creating new variables from the existing data that might be helpful for the machine learning algorithm. For example, if we're predicting equipment failure, we might create a new variable that represents the change in temperature over the last hour.

3. **Splitting the Data**: The data is typically divided into two or three subsets: a training set, a validation set, and a test set. The training set is used to train the model, the validation set is used to tune model parameters and prevent overfitting, and the test set is used to evaluate the model's performance on unseen data.

4. **Model Selection and Training**: A machine learning algorithm (or multiple algorithms) is chosen and trained on the training data. This involves feeding the input data (features) into the algorithm and adjusting the model's internal parameters based on the output (predictions) and the actual values (labels). The goal is to adjust the parameters so that the model makes as few errors as possible.

5. **Model Validation and Tuning**: The model's performance is evaluated on the validation set. If the model's performance is unsatisfactory, hyperparameters can be tuned (like the learning rate in a neural network or the depth of a decision tree). This step might also involve techniques to avoid overfitting, like regularization.

6. **Model Evaluation**: Finally, the model's performance is evaluated on the test set. This provides an unbiased assessment of the model's performance, as the test set was not used during training or validation.

7. **Iterate**: The above process is iterative. You should go back and perform additional preprocessing, feature engineering, or try different models based on the results you're seeing.

This is a simplified explanation. The exact process will depend on the machine learning type (supervised, unsupervised, reinforcement), the specific algorithm, and the problem being solved.

Predictive Modeling involves creating, testing, and validating a model to predict the probability of an outcome best. In predictive maintenance, predictive models take data from various sources to forecast potential failures. For example, a model might analyze the relationship between a machine's operating temperature and failure rates and then use this relationship to predict future failures based on observed temperature data.

Analyzing the relationship between a machine's operating temperature and failure rates involves statistical analysis and may require the application of machine learning techniques. See Figure 9 and the steps below to illustrate several analysis types used for predictive modeling.

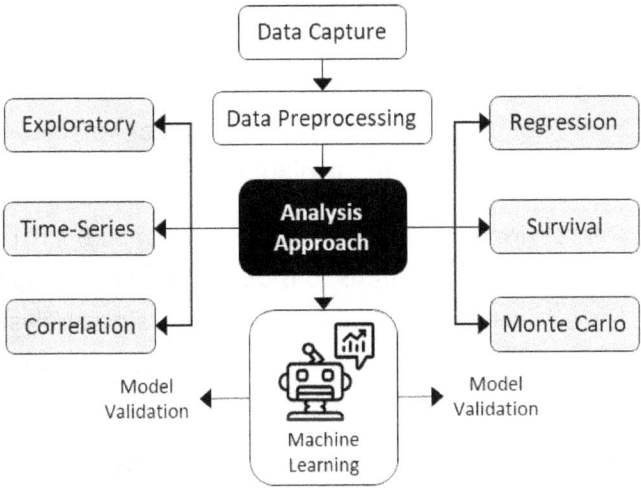

Figure 9: Steps of Predictive Modeling

1. **Data Collection**: Collect historical data on the machine's operating temperature and instances of failure. The more data you have, the more reliable your analysis will be.

2. **Data Cleaning and Preprocessing**: Clean the collected data and handle missing values and outliers if necessary. Transform the data into a suitable format for analysis. For instance, categorize temperatures into ranges (e.g., low, medium, high) or normalize the temperature data.

3. **Exploratory Data Analysis (EDA)**: Perform initial analysis to understand the data's characteristics. This might involve visualizing the data with scatter plots, box plots, or histograms. Plotting failure rates against temperature can give an initial understanding of the relationship between these two variables.

4. **Correlation Analysis**: Calculate the correlation between temperature and failure rates. A positive correlation indicates that the failure rate also increases as temperature increases, while a negative correlation indicates the opposite. Correlation coefficients range from -1 to +1, where +1 is a total positive linear correlation, 0 is no linear correlation, and -1 is a total negative linear correlation.

5. **Regression Analysis**: If a linear relationship exists, you might use regression analysis to model the relationship between temperature and failure rates. Regression analysis could provide a mathematical model to predict failure rate based on temperature.

6. **Time-Series Analysis**: Specific time-series analysis techniques might be useful if the data is time-series (collected over time). Time series analysis is a specific way of analyzing a sequence of data points collected over time. In time series analysis, data points are recorded at consistent intervals over a set length

rather than just recording the data points intermittently or randomly.

7. **Monte Carlo Simulation**: The Monte Carlo simulation is a mathematical technique that predicts possible outcomes of an uncertain event. Computer programs use this method to analyze past data and predict future outcomes based on a choice of action. It predicts outcomes of an uncertain event (e.g., equipment failure) using probability.

8. **Survival Analysis**: This statistical approach deals with the time until an event happens – in this case, the time until a machine fails. This analysis aims to simultaneously evaluate the effect of several factors on survival.

9. **Machine Learning**: More complex relationships might require machine learning techniques. For instance, if the relationship between temperature and failure is non-linear or affected by other variables, techniques like decision trees, random forests, or neural networks might be more appropriate.

10. **Model Validation**: Whatever model you create must be validated to ensure accuracy. This might involve techniques like cross-validation, where you test the model on a subset of your data to see how well it performs.

Remember to interpret the results carefully and consider other variables affecting machine failure. A relationship between temperature and failure rate does not necessarily mean that high temperature causes failure – it might be a symptom rather than a cause.

Anomaly Detection is a technique used to identify patterns in each dataset that do not conform to an established normal behavior. These anomalies often translate to critical, actionable information in a predictive maintenance context. A machine's typical operation creates a normal data pattern, and any significant deviation from this pattern (an anomaly) might indicate a problem.

Creating a pattern of regular operation in predictive maintenance involves understanding what normal behavior looks like for a machine or a system. This normal behavior forms a baseline against which new observations can be compared to detect anomalies or deviations. Here's how this is typically done:

1. **Data Collection**: The first step is collecting historical or real-time data from the system.

2. **Data Preprocessing**: The collected data is then cleaned and normalized.

3. **Baseline Creation**: Once the data is preprocessed, it is used to establish a baseline or "normal" operation pattern. This could involve calculating the mean and standard deviation of the collected measurements or using more complex techniques to model the machine's normal behavior.

4. **Anomaly Detection**: New data from the machine is continuously compared against this baseline. If the new data significantly deviates from the baseline, it's considered an anomaly. Anomaly detection could be as simple as checking if a new measurement is within a specific range of the mean, or it could involve more complex techniques like machine learning algorithms.

5. **Actionable Insights**: Once an anomaly is detected, it's essential to determine whether it indicates a potential issue. This might involve consulting with domain experts or further analysis. If the anomaly is considered significant, a maintenance action may be triggered to prevent a possible failure.

6. **Continuous Monitoring and Model Updating**: The machine's performance continues to be monitored, and the baseline model is updated with new data to ensure accuracy.

These steps help predictive maintenance by catching potential issues before they result in machine failure, saving time and money, and preventing possible disasters. See Figure 10.

Forecasting is making predictions based on past and present data. In the realm of predictive maintenance, forecasting might involve predicting when a piece of equipment is likely to fail based on historical data and current operational data. This predictive capability allows for proactive maintenance and reduces the likelihood of unexpected equipment failures. To examine the forecasting process further, it helps to view a particular algorithm, such as a neural network.

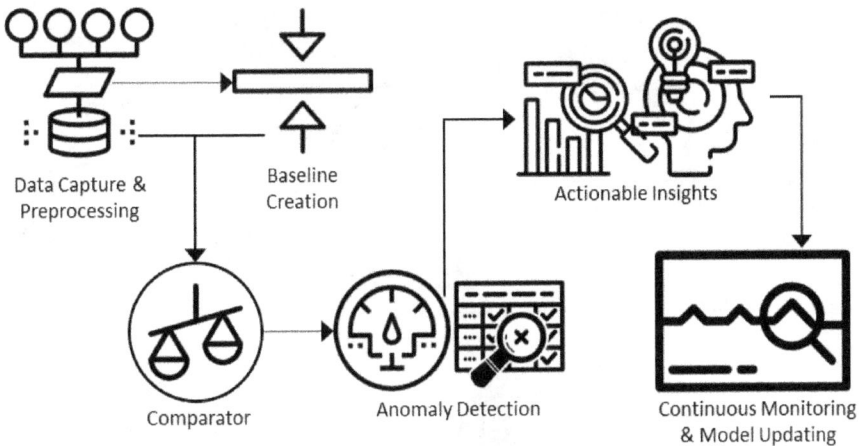

Data Capture & Preprocessing

Baseline Creation

Actionable Insights

Comparator

Anomaly Detection

Continuous Monitoring & Model Updating

Figure 10: Anomaly Detection for PdM

Neural networks are a type of machine learning algorithm that's particularly well-suited for complex pattern recognition tasks, making them an excellent tool for forecasting in predictive maintenance. A general overview of how neural networks are used in this context is shown in Figure 11 and the items below.

1. **Data Collection and Preprocessing**: As with any machine learning application, the first steps involve gathering relevant data (e.g., sensor readings, operational conditions, maintenance logs, failure instances) and preparing it for the neural network.

2. **Feature Extraction**: Neural networks, especially deep learning models, are known for their ability to perform automatic feature extraction. They can identify relevant patterns or features in the input data that predict equipment failure. This can often lead to better performance than manually engineered features.

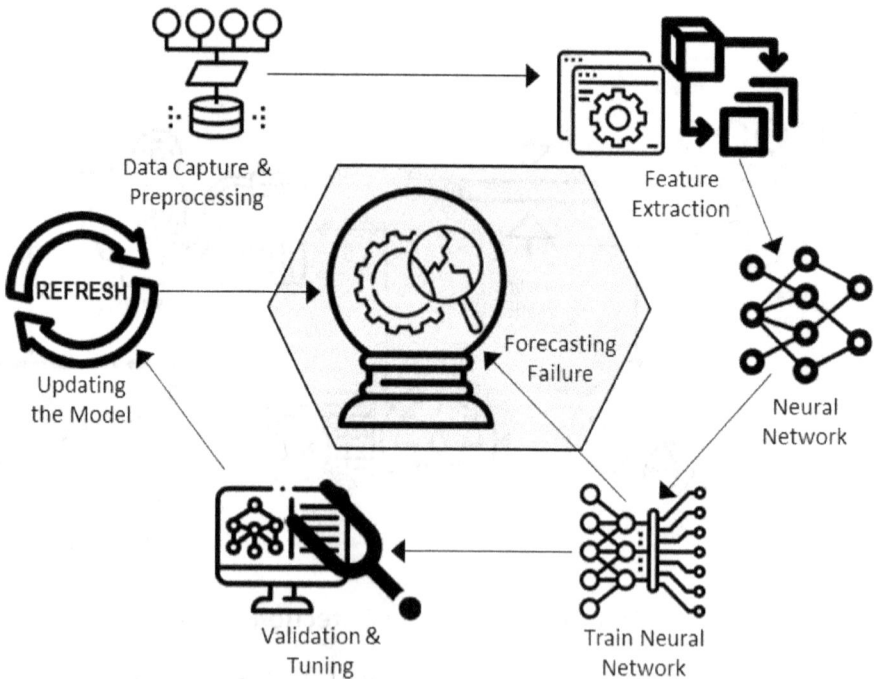

Figure 11: Forecasting Failure Utilizing Neural Network

3. **Building the Neural Network**: A neural network is designed with input layers that correspond to the features (data points), hidden layers that process and extract patterns from the data, and an output layer that makes the final prediction (e.g., time to failure, probability of failure within a certain period, etc.).

4. **Training the Network**: During training, the network learns by adjusting the weights and biases in response to the input data and the associated outcome. This is typically done using a method called backpropagation and a suitable optimization algorithm, like stochastic gradient descent.

5. **Validation and Tuning**: The network's performance is tested on a separate validation dataset, which it has yet to see during training. If the model performs well, it's ready to be deployed. Otherwise, its architecture or hyperparameters may need to be adjusted, and it might need to be retrained.

6. **Forecasting**: Once the neural network is trained and validated, it can forecast future failures or maintenance needs. It does this by analyzing the current and historical operational data, identifying the learned patterns that indicate a failure, and making predictions accordingly.

7. **Updating the Model**: One advantage of using neural networks for predictive maintenance is that they can continue to learn as new data becomes available. This allows the model to adapt to changes in equipment behavior over time.

Using neural networks in predictive maintenance forecasting can lead to more accurate and timely predictions of equipment failure. This, in turn, can result in significant cost savings, increased equipment uptime, and improved operational efficiency.

These key concepts form the foundation of predictive maintenance, allowing for a data-driven, proactive approach to equipment upkeep that minimizes downtime and optimizes efficiency. The depth of knowledge in these areas determines the efficacy of a predictive maintenance strategy.

Part II: The Mechanics of Predictive Maintenance

Chapter 4: Essential Elements of Predictive Maintenance

Predictive maintenance is a symbiotic interplay between hardware and software. It hinges upon a robust architecture of sensors, IoT devices, and machinery for data collection while utilizing advanced software tools to perform data analytics and apply artificial intelligence and machine learning techniques. These elements form an intricate web that monitors, predicts, and prevents equipment failures. See Figure 12 and the items listed below.

Hardware in predictive maintenance comprises the physical components that work synergistically to monitor the health of various equipment and systems.

Sensors play a critical role in gathering valuable data such as temperature, pressure, vibration, humidity, and acoustic signals, which can indicate anomalies or changes in machine performance.

IoT devices, or Internet of Things devices, are typically embedded with sensors and software to collect and transmit data over the internet. These devices form a network of connected objects that continuously gather, send, and receive data, enabling real-time monitoring and reporting of equipment status.

Equipment is the primary subject of predictive maintenance. This includes machinery, components, and systems whose performance is continually tracked. The nature of the equipment can vary widely based on the industry, ranging from heavy machinery in manufacturing plants to HVAC systems in commercial buildings.

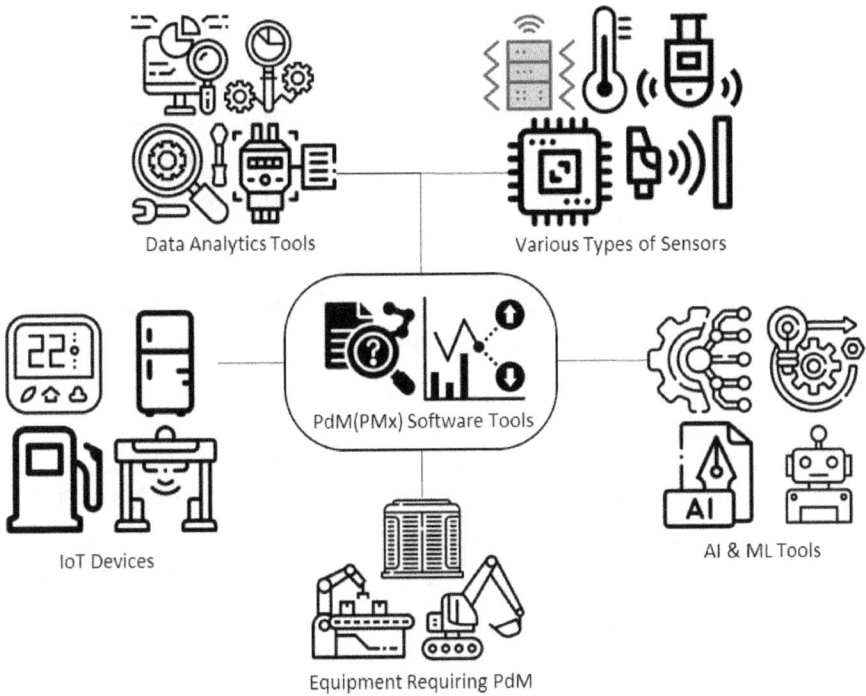

Figure 12: PdM (PMx) Equipment & Hardware/Software

On the other side of the equation is **Software**, the digital backbone of predictive maintenance, which processes, analyzes, and makes sense of the wealth of data the hardware collects.

Data Analytics tools are used to clean, organize, and analyze the collected data. These tools can handle large datasets and perform complex computations to extract valuable insights.

AI and ML Tools play an instrumental role in predictive maintenance. Machine learning algorithms learn from historical and real-time data patterns to predict future equipment failures. They can be trained to recognize patterns associated with specific equipment faults and predict their occurrence. AI, in turn, can assist in making data-driven decisions, automating the scheduling of maintenance tasks, and even predicting the impact of these tasks on operations.

In essence, the effective deployment of predictive maintenance depends on harmoniously integrating the physical and digital worlds. By leveraging the strengths of both hardware and software elements, businesses can create a highly responsive and proactive maintenance environment that substantially boosts their operational efficiency and productivity.

Predictive maintenance involves several steps, from sensor installation and data collection to final decision-making. A detailed description of how predictive maintenance is executed is shown in Figure 13, and the steps are listed below.

1. **Sensor Installation:** The first step is to install appropriate sensors on the equipment to monitor various parameters. The selection of sensors depends on the specific equipment and the data type needed. Common sensors include vibration sensors, temperature sensors, pressure sensors, flow meters, acoustic sensors, and more. These sensors are strategically placed to capture relevant data on the equipment's health and performance.

2. **Triggering Events:** Once the sensors are installed, triggers or thresholds are set to determine when data collection and analysis should occur. These triggers are typically defined based on pre-established rules or algorithms. For example, a trigger could be a specific vibration level exceeding a threshold, a sudden temperature rise, or a deviation from the normal operating parameters. When a triggering event occurs, it initiates the data collection process.

3. **Data Collection:** When a trigger event is detected, the sensors collect data from the equipment. This data could include sensor readings, operational parameters, environmental conditions, or other relevant information. The data collection process can be continuous, periodic, or event-based,

depending on the specific requirements and capabilities of the system.

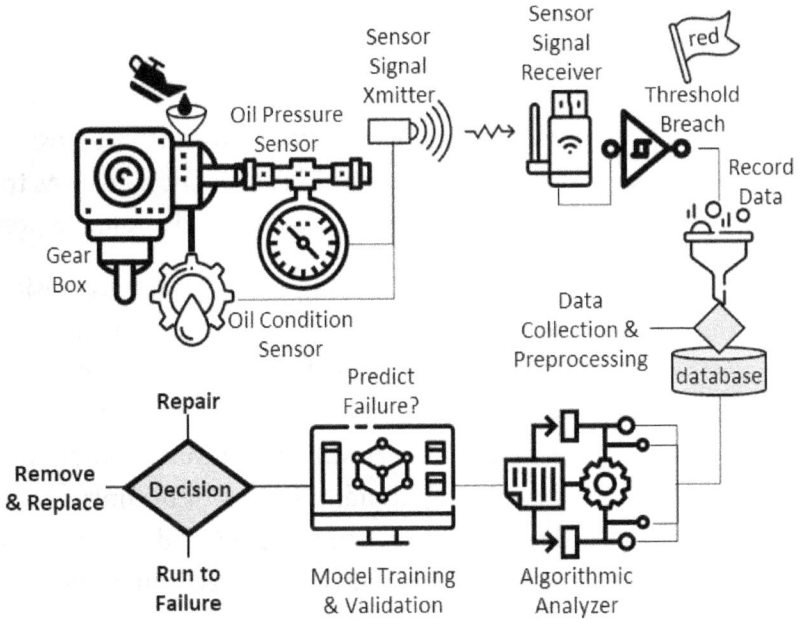

Figure 13: Predictive Maintenance Process

4. **Data Preprocessing:** The collected data may need to undergo preprocessing steps to ensure its quality and compatibility for further analysis. This step involves cleaning the data by removing outliers, handling missing values, and normalizing the data to a consistent format. Additionally, data may need to be synchronized to align different sensor readings and timestamps.

5. **Algorithmic Analysis:** Once the data is preprocessed, it is fed into algorithms for analysis. Various machine learning and statistical techniques extract meaningful insights and predict potential failures. The choice of algorithms depends on the specific goals and requirements of the predictive maintenance system. Commonly used algorithms include regression

models, classification algorithms, anomaly detection algorithms, and time-series analysis techniques.

6. **Model Training and Validation:** The predictive models are trained using historical data that includes normal operating conditions and known failure instances. The trained models are then validated using separate datasets to assess their accuracy and performance. This validation ensures the models can generalize well to new data and make reliable predictions.

7. **Decision Making:** Based on the predictions and insights generated by the predictive models, decisions regarding maintenance actions are made. The decisions could include scheduling preventive maintenance, replacing a component, adjusting operational parameters, or taking other appropriate action to prevent equipment failures or optimize maintenance efforts. These decisions are typically made by maintenance teams or integrated into an automated maintenance system.

It's important to note that the execution of predictive maintenance can vary based on the specific industry, equipment, and available resources. However, the general process involves sensor installation, trigger events, data collection, data preprocessing, algorithmic analysis, modeling, model training, validation, and decision-making steps to optimize equipment availability and minimize downtime.

Chapter 5: Data Collection and Management

Predictive maintenance is fundamentally driven by data, making its collection and management crucial to the success of this strategy. The data collected can vary, ranging from sensor readings to operational logs, and the accuracy and reliability of this data form the bedrock of predictive insights. It's essential to understand best practices for data gathering and the nuances of data quality, cleaning, and management to implement effective predictive maintenance.

Data Gathering forms the initial phase in the predictive maintenance journey. This involves systematically collecting data from various sources such as equipment sensors, IoT devices, operation and maintenance logs, and even external factors like weather conditions for specific industries. The frequency of data collection should align with the objectives of the predictive maintenance program, the nature of the equipment, and the type of data being collected. Collecting data at an unnecessarily high frequency can lead to data overload and increased storage needs, while infrequent collection may miss important events or trends.

Best practices for data gathering involve ensuring that the correct sensors are installed at appropriate locations to provide relevant and comprehensive data. Maintaining the hardware regularly is also important to avoid inaccurate readings due to faulty sensors. Proper networking and connectivity are equally essential to ensure the data is reliably transmitted to the central system or cloud for further analysis.

Data Capture Technologies play a crucial role in predictive maintenance by collecting relevant information from various sources to monitor the condition of equipment and predict potential failures. Some common data capture technologies used in predictive maintenance are listed below. See Figure 14.

1. **Sensors:** Sensors are widely used to capture real-time data on parameters such as temperature, pressure, vibration, humidity, and more. They are attached to equipment or machinery and provide continuous measurements, which can be analyzed to detect anomalies or patterns indicating potential failures.

2. **Internet of Things (IoT) Devices:** IoT devices enable data collection from a network of interconnected sensors and equipment. These devices can transmit data to a central system for analysis, allowing for remote monitoring and predictive maintenance. IoT devices can include smart sensors, gateways, and edge computing devices.

3. **Machine Vision:** Machine vision systems use cameras and image processing algorithms to capture and analyze visual data from equipment. They can detect defects, measure dimensions, and monitor the condition of components. Machine vision can be used for predictive maintenance in industries such as manufacturing and automotive.

4. **Digital Twins:** Digital twins are virtual replicas of physical assets or systems. They combine real-time data from sensors, historical data, and models to simulate the behavior and condition of the physical asset. By capturing and analyzing data from digital twins, predictive maintenance can be performed to identify potential failures and optimize maintenance schedules.

5. **Condition Monitoring Systems:** Condition monitoring systems employ a combination of sensors, data acquisition units, and software to monitor the health of equipment continuously. These systems capture and analyze data on various parameters to detect early signs of degradation or impending failures.

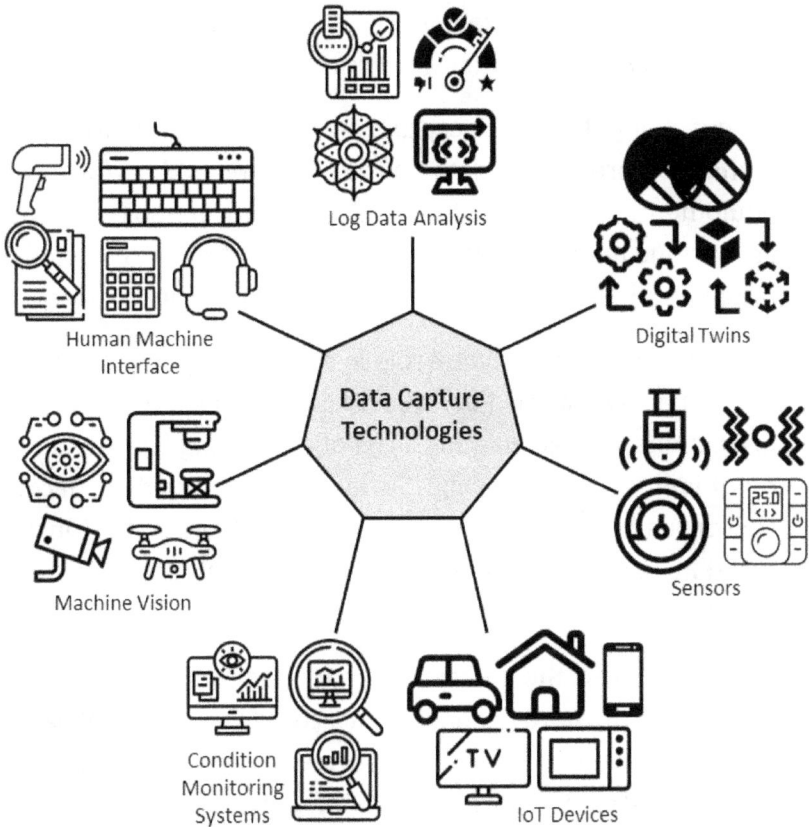

Figure 14: Data Capture Technologies

6. **Log Data Analysis:** Log data captures information about system events, error messages, and performance metrics. Analyzing log data using techniques such as log mining and anomaly detection can provide insights into the condition and behavior of equipment. It can help identify patterns indicating potential failures or degradation.

7. **Human-Machine Interface (HMI):** HMIs allow operators or maintenance personnel to interact with equipment and capture relevant data manually. This can include entering observations, recording maintenance activities, or inputting data during inspections. Although manual data capture may

be less automated, it can still be valuable for predictive maintenance when combined with other data sources.

By leveraging these data capture technologies, organizations can gather comprehensive data on the condition and performance of equipment. This data is then analyzed using various techniques such as machine learning, statistical analysis, and pattern recognition to predict potential failures and optimize maintenance strategies.

Data Quality plays a pivotal role in the effectiveness of a predictive maintenance system. Poor quality data can lead to inaccurate predictions, missed warning signs of impending failures, and misguided decision-making. Several factors, including inaccurate sensor readings, missing data, inconsistent data collection methods, and data corruption during storage or transmission, can impact data quality.

Data Cleaning is the process of identifying and correcting or removing data errors to improve quality. This could involve handling missing values, correcting inconsistent entries, and removing outliers or noise that could skew the data analysis. Effective data cleaning can significantly enhance the accuracy of the predictive models and the value of the insights derived.

Data Management involves the storage, organization, and security of the data. With the large volumes of data generated in a predictive maintenance setup, efficient data management becomes crucial. It's essential to have robust data storage solutions that can handle the scale of data while ensuring quick and easy access. Effective data management also involves organizing the data to facilitate efficient retrieval and analysis. Moreover, stringent measures should be in place to ensure data security and compliance with relevant regulations.

In conclusion, effective data collection and management are at the heart of predictive maintenance. Ensuring the right data is collected, maintaining high data quality, conducting thorough data cleaning, and having robust data management practices can significantly boost the success of a predictive maintenance program.

Chapter 6: Data Analysis and Interpretation

Once the data has been collected and cleaned, the next step in the predictive maintenance process is data analysis and interpretation. Here, different types of analysis are performed to extract meaningful insights from the data, and algorithms and models are developed to predict future equipment failures.

Descriptive Analysis is the first level of data analysis, which provides insights into the past. It uses data aggregation and data mining techniques to provide insight into the past and answer: "What has happened?" In a predictive maintenance context, this could involve summarizing historical machine performance data or analyzing past instances of equipment failure to identify patterns or trends.

Diagnostic Analysis takes the insights gained from descriptive analysis further to answer the question: "Why did it happen?" This involves delving deeper into the data to understand the root cause of past events. For example, diagnostic analysis in predictive maintenance might involve examining why a piece of equipment failed or identifying the factors contributing to increased machine downtime.

Predictive Analysis uses statistical models and forecasting techniques to understand the future. It answers the question: "What could happen?" This is the heart of predictive maintenance, as predictive models are developed to anticipate equipment failures based on patterns identified in the data. Machine learning algorithms play a crucial role here, learning from historical data to predict future events.

Prescriptive Analysis goes beyond predictive analysis to recommend actions that take advantage of the predictions. It answers the question: "What should we do?" In predictive maintenance, prescriptive analysis might involve suggesting the optimal maintenance schedule to prevent predicted equipment failures or recommending adjustments to machine operation to prolong equipment life.

Developing **Algorithms and Models** is a central part of predictive maintenance. Once the data has been collected, cleaned, and analyzed, machine learning algorithms are trained on this data to create predictive models. These models use the patterns and relationships learned from the historical data to predict future equipment failures. It's essential to validate and regularly update these models to ensure they provide accurate predictions as new data is collected and the equipment and operational conditions change over time.

Predictive Maintenance Algorithms involve using various machine learning and statistical methods to predict when a device or system might fail. By forecasting these failures, maintenance can be carried out in advance to prevent them, saving both time and resources. Some of the most used predictive maintenance algorithms for predicting failure are listed below.

1. **Regression Algorithms:** These are predictive modeling techniques that aim to understand the relationship between a dependent variable (e.g., time until machine failure) and one or more independent variables (e.g., machine usage time, heat generated, etc.). Examples include Linear Regression, Polynomial Regression, and Support Vector Regression. See Figure 15.

2. **Classification Algorithms:** These algorithms predict the category or class of a given equipment condition. Examples

include Logistic Regression, Decision Trees, Random Forests, and Gradient Boosting Machines. These models can categorize a machine's condition as normal, abnormal, or failure imminent.

3. **Time Series Analysis and Forecasting:** These are methods used to analyze time-based data, considering trends, cyclic patterns, and seasonal effects. Algorithms like ARIMA (Auto Regressive Integrated Moving Average), LSTM (Long Short-Term Memory), and Prophet can be used for this purpose. This is often used when data is collected over time, like machine sensor data.

4. **Survival Analysis:** Also known as event history analysis or reliability analysis in engineering, these statistical methods estimate the time until an event (e.g., machine failure) occurs. The Cox Proportional Hazards model is a standard method used here.

5. **Neural Networks:** These deep learning algorithms can recognize patterns and trends in vast amounts of data, making them a powerful tool for predictive maintenance. A neural network can be utilized for predictive maintenance by learning patterns and relationships from historical data to predict future equipment failures or maintenance needs.

6. **Anomaly Detection Algorithms:** These algorithms identify outliers in the data. In predictive maintenance, they could help detect unusual behavior of a machine, which may indicate a problem. Examples include Isolation Forest, One-Class SVM, and Autoencoders in deep learning.

7. **Reinforcement Learning:** In some instances, reinforcement learning can be used to define and implement a policy for maintenance. The algorithm learns the best policy by exploring different maintenance actions and their outcomes.

In conclusion, data analysis and interpretation in predictive maintenance involve extracting insights from historical data, understanding the reasons for past equipment failures, predicting future failures, and prescribing actions to prevent these predicted failures. The development, validation, and continuous improvement of predictive models are central to this process, enabling proactive and data-driven maintenance strategies. Remember that selecting the right algorithm(s) depends on many factors, including the type and quantity of data available, the specific problem you're trying to solve, the computational resources at your disposal, and the level of accuracy required.

As seen above, there are many algorithms to choose from when approaching predictive maintenance. To understand further, let's examine two types of algorithms: (1) regression algorithm and anomaly detection algorithm.

Using a regression algorithm for predictive maintenance involves several steps. See Figure 15 and the steps below for a general outline of the process.

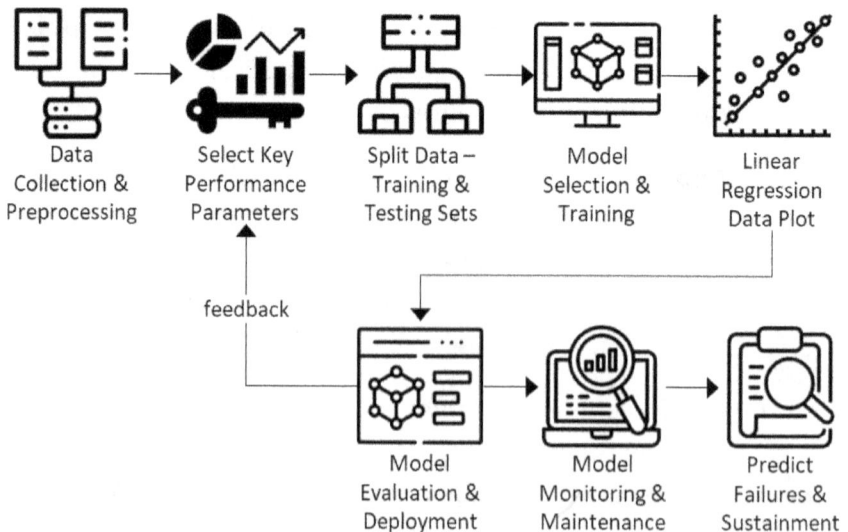

Data Collection & Preprocessing → Select Key Performance Parameters → Split Data – Training & Testing Sets → Model Selection & Training → Linear Regression Data Plot

feedback

Model Evaluation & Deployment → Model Monitoring & Maintenance → Predict Failures & Sustainment

Figure 15: Regression Algorithm Methodology

1. **Data Collection:** Gather historical data related to your maintenance activities and equipment. This data should include relevant features such as sensor readings, operating conditions, maintenance records, failure events, and other information that could impact equipment performance.

2. **Data Preprocessing:** Clean the data and prepare it for analysis. This step may involve handling missing values, removing outliers, normalizing or scaling features, and encoding categorical variables.

3. **Feature Selection/Engineering:** Select the most relevant features for your predictive maintenance model. You can use domain knowledge or techniques like correlation analysis to identify the features most impacting equipment failure. Additionally, you can create new features by combining or transforming existing ones to capture essential patterns.

4. **Split the Data:** Divide your dataset into training and testing sets. The training set will be used to train the regression algorithm, while the testing set will be used to evaluate its performance.

5. **Model Selection:** Choose a regression algorithm suitable for your predictive maintenance problem. Common choices include linear regression, logistic regression, support vector regression, decision trees, random forests, or gradient-boosting algorithms. The selection depends on the nature of your data and the type of regression problem (e.g., predicting remaining useful life, failure probability, or time to failure).

6. **Model Training:** Train the regression model using the training data. The algorithm will learn the patterns and relationships

between the input features and the target variable (e.g., remaining useful life or maintenance cost).

7. **Model Evaluation:** Assess the performance of your trained model using the testing data. Common evaluation metrics for regression models include mean squared error (MSE), root mean squared error (RMSE), mean absolute error (MAE), or R-squared (coefficient of determination). Evaluate the model's ability to accurately predict the target variable and consider if the results meet your requirements.

8. **Model Deployment:** Once satisfied with the model's performance, deploy it in your predictive maintenance system. This involves integrating the model into your production environment and using it to make predictions on new data.

9. **Monitoring and Maintenance:** Continuously monitor the performance of your predictive maintenance model in the real-world setting. Monitor the model's predictions and compare them to actual maintenance events to ensure they remain accurate and reliable. If necessary, retrain or update the model periodically as new data becomes available.

Remember that the specifics of implementing a regression algorithm for predictive maintenance may vary depending on your specific use case and data characteristics. Tailoring the approach to your needs and iterating as you gain more insights and refine your model is essential.

Using an anomaly detection algorithm for predictive maintenance can be highly beneficial in identifying potential faults or failures in machinery or equipment. A step-by-step guide on how to utilize an anomaly detection algorithm for predictive maintenance is provided below. In addition, see Figure 16.

1. **Gather historical data:** Collect substantial historical data related to the machinery or equipment you want to monitor. This data should include sensor readings, operating conditions, maintenance records, and other relevant information.

2. **Preprocess the data:** Clean the data by removing any outliers or inconsistencies. Perform data normalization or standardization to ensure all variables are on a consistent scale. Additionally, handle missing data points through imputation techniques such as interpolation or mean substitution.

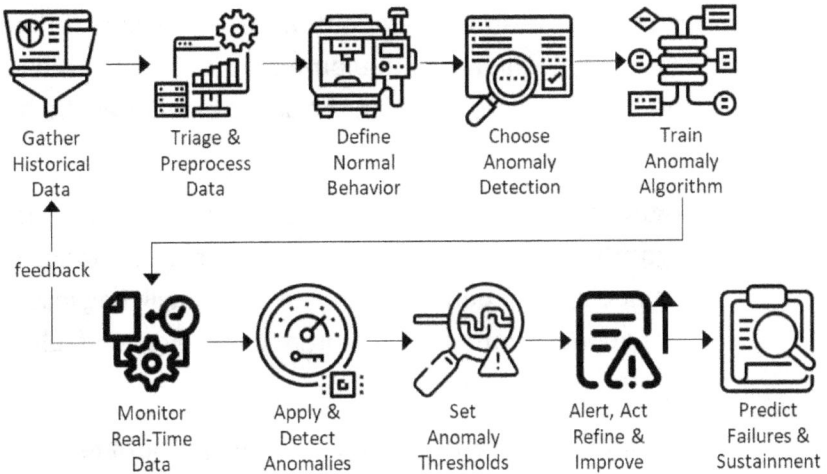

Figure 16: Anomaly Detection Algorithm

3. **Define normal behavior:** Identify the machinery or equipment's baseline or normal behavior based on the historical data. This can be achieved by analyzing statistical measures such as mean, standard deviation, or percentiles. You may also consider using domain knowledge or expert input to define normal operating conditions.

4. **Choose an anomaly detection algorithm:** Select an appropriate one that suits your specific needs and the characteristics of your data. Commonly used algorithms for anomaly detection include statistical methods (e.g., Z-score, Gaussian distribution), machine learning algorithms (e.g., isolation forest, one-class SVM), and deep learning techniques (e.g., autoencoders, variational autoencoders).

5. **Train the algorithm:** Use the preprocessed historical data to train the selected anomaly detection algorithm. Depending on the algorithm, you may need to adjust specific parameters or hyperparameters to achieve optimal performance.

6. **Monitor real-time data:** Continuously collect real-time data from the machinery or equipment as it operates. Ensure that the data is preprocessed in the same manner as the historical data.

7. **Detect anomalies:** Apply the trained anomaly detection algorithm to the real-time data to identify deviations from the defined normal behavior. The algorithm will flag instances that appear anomalous or potentially indicate a fault or failure.

8. **Set anomaly thresholds:** Determine appropriate thresholds for triggering an anomaly alert or maintenance action. These thresholds can be based on statistical measures or determined through experimentation and validation.

9. **Alert and act:** If the anomaly detection algorithm identifies an anomaly beyond the set threshold, generate an alert or notification to the relevant personnel responsible for maintenance. Promptly investigate the issue and take appropriate action, such as scheduling maintenance, repairs, or inspections.

10. **Refine and improve:** Continuously evaluate the performance of the anomaly detection algorithm and refine it over time. Incorporate feedback from maintenance actions and monitor false positives or negatives to improve the accuracy and effectiveness of the predictive maintenance system.

Remember that deploying an anomaly detection algorithm for predictive maintenance is an iterative process that requires ongoing monitoring, refinement, and collaboration between data scientists, maintenance personnel, and domain experts to achieve optimal results.

Chapter 7: Predictive Maintenance and Calculations

It is helpful to provide an example of predictive maintenance in action. For purposes of clarity and simplicity, the following data is provided. Remember we are keeping this example at five data inputs; there may be additional variables, e.g., type of maintenance, preventive maintenance schedule, engine oil temperature, engine rpm, etc.

1. Asset = delivery truck
2. Operational hours = 10,000
3. Mean time between failures = 90 days
4. Mean time to repair = 200 hours
5. Last failure occurred = 9000 hours

First, we need to convert MTBF from days to hours because we need the same units for our calculations. There are 24 hours in a day, so:

MTBF = 90 days * 24 hours/day = 2160 hours

Then, we can calculate the expected number of failures. The number of failures is given by the total time divided by the MTBF.

Number of Failures = Total Hours / MTBF

Substituting the given values:

Number of Failures = 10,000 hours / 2160 hours = 4.63

So, if the truck is operated for 10,000 hours, you might expect approximately 4.63 failures, assuming the failures are independent and identically distributed.

Again, note that the number of failures is a statistical average. In practice, the truck won't experience 4.63 failures; it will experience 4 or 5 or some other whole number of failures. This is an expectation value based on the statistical analysis. See Figure 17 for a depiction of this example.

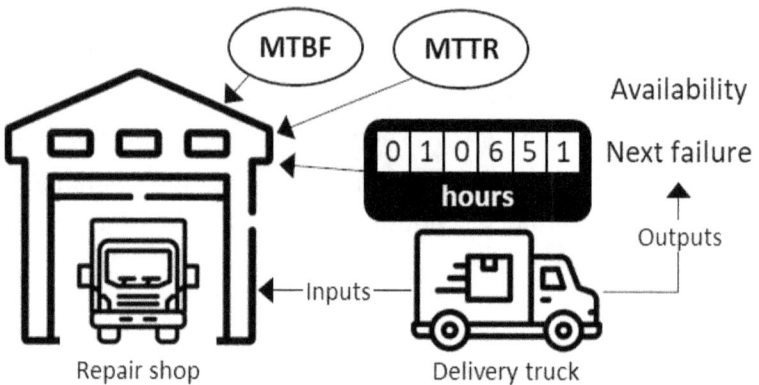

Figure 17: PdM Calculations Example

In terms of availability (which is the proportion of time the truck is expected to be operational), we use the formula:

Availability = MTBF / (MTBF + MTTR)

Now, plug these values into the availability formula:

Availability = 2160 hours / (2160 hours + 200 hours) = 2160 / 2360 = 0.9153 = 91.53%

This means we can expect the truck to be operational about 91.53% of the time.

Finally, assuming failures occur independently and are identically distributed, the subsequent failure can be forecasted as described below.

We can estimate the subsequent failure using the Mean Time Between Failures (MTBF) based on the given data. In this case, MTBF is 2160 hours (which we calculated from the given 90 days), meaning, on average, we expect a failure every 2160 hours.

Given that the last failure was at 9000 hours, we might expect the next failure around:

9000 hours (time of the last failure) + 2160 hours (MTBF) = 11,160 hours

However, remember that the MTBF is an average, and the actual time to the subsequent failure can vary. The prediction is a statistical one and does not account for many factors that can influence when a failure occurs, such as changes in usage patterns, environmental conditions, quality of maintenance, etc. Therefore, while the estimate provides a guideline, it should be used with ongoing monitoring and maintenance practices.

It's also worth noting that statistical estimates like this are most reliable when used to understand performance over more extended periods or across multiple vehicles rather than predicting the precise timing of a single event.

Another example of an HVAC system is provided in Appendix C.

AI (Artificial Intelligence) and Machine Learning techniques can be used to predict future failures in machinery and equipment, a field often referred to as predictive maintenance. However, the accuracy of these predictions largely depends on the quality and quantity of the data available.

A few points to note are provided below. The data would be captured over longer intervals and consist of several assets to provide more accuracy in the predictive maintenance models.

1. **Historical Data:** To train a model capable of predicting machinery failures, we need historical data about previous breakdowns, maintenance, operational conditions, and other potentially influential factors. This data is used to identify patterns or signs that often precede a failure.

2. **Features:** The success of the prediction will depend heavily on what features (input variables) are available for prediction. This could include usage statistics, environmental conditions, past maintenance, and any other data that could impact the machinery's reliability.

3. **Models:** Different machine learning models can be applied, including regression models, decision trees, and neural networks. The choice of model may depend on the nature of the problem and the data available.

4. **Continuous Monitoring:** Real-time monitoring of equipment is also often beneficial. Sensor data can feed into AI models in real time, allowing potential failures to be detected and predicted more accurately and earlier.

5. **Evaluation & Adjustment:** Once a predictive model is in place, it should be regularly evaluated and adjusted as necessary based on its performance over time.

It's important to note that while AI can make accurate predictions, it is not infallible. Any predictions should be used as a guide and complemented with expert human judgment.

Part III: Implementing Predictive Maintenance

Chapter 8: Developing a Predictive Maintenance Plan

Building a predictive maintenance program is not an overnight task but a journey that requires careful planning and execution. A well-devised plan that accounts for the organization's goals and objectives, the current capabilities and resources, and a clear roadmap for implementation and evolution can significantly increase the odds of success. This task is the very first one that needs to be completed. Formulate a plan, determine who the champions are, identify your resources, and research the topics and ideas. There are several excellent sources; a few of them are listed in the bibliography and references at the end of this book. A detailed journey to plan preparation is provided in the steps below and Figure 18.

Identifying Goals and Objectives is the initial step in developing a predictive maintenance plan. Clear, well-defined goals guide the plan and keep all stakeholders aligned. These goals include reducing equipment downtime, increasing the operational life of machinery, reducing maintenance costs, or improving overall operational efficiency. Along with overarching goals, specific, measurable objectives should also be set to track the progress of the predictive maintenance program.

Assessing Current Capabilities and Resources is crucial to understanding the starting point of the predictive maintenance journey. This involves evaluating the existing maintenance practices, the condition and criticality of the equipment, the available data and how it's currently used, the existing IT infrastructure, and the skills and competencies of the staff. This assessment can highlight the strengths to build on, the gaps to fill, and the challenges to prepare for in implementing predictive maintenance.

Creating a Roadmap involves planning the steps to move from the current state to the desired state, as defined by the goals and objectives. This roadmap should consider the capabilities and resources assessment results, and outline the actions needed to implement predictive maintenance effectively. These actions could involve installing or upgrading sensors and IoT devices, acquiring or developing the necessary software tools, training, hiring staff with the needed skills, and defining the data collection, analysis, and decision-making processes. See Figure 18.

The roadmap should also plan for the ongoing evolution and improvement of the predictive maintenance program.

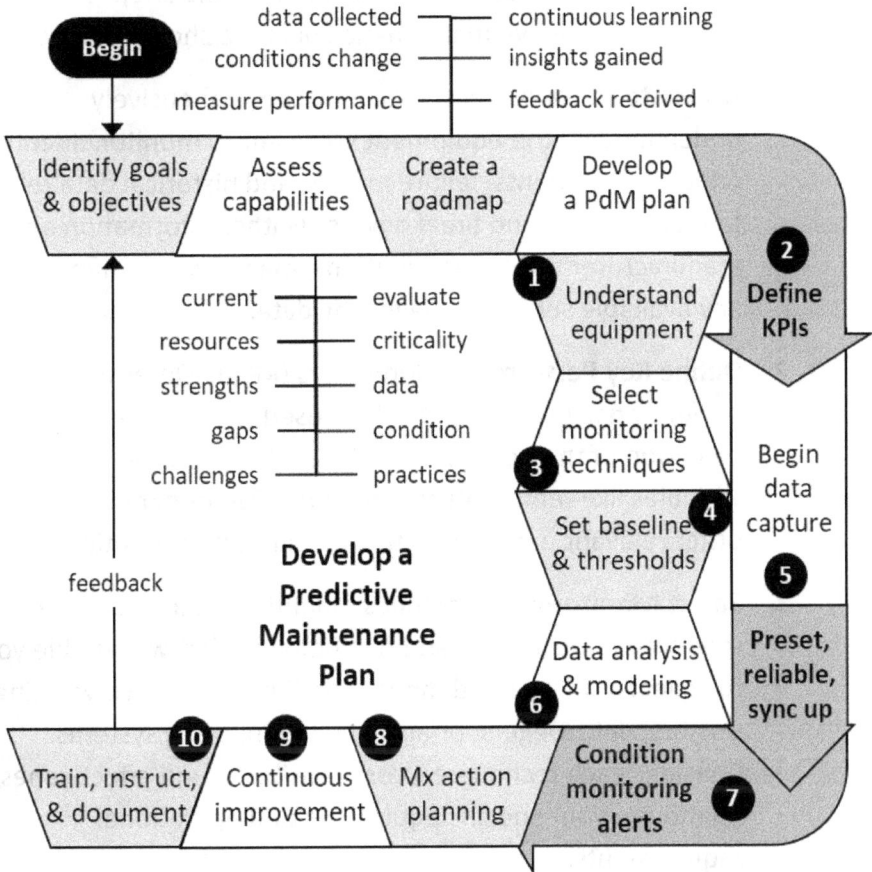

Figure 18: Create a Predictive Maintenance Plan Strategy

This includes regular reviews of the program's performance against the set objectives, updates to the predictive models as more data is collected and conditions change, and continuous learning and adaptation based on the insights gained and the feedback received.

Developing a Predictive Maintenance Plan involves a systematic approach to anticipate equipment failures or malfunctions before they occur. By implementing predictive maintenance strategies, businesses can reduce downtime, optimize maintenance schedules, and increase overall operational efficiency. The steps to develop a predictive maintenance plan are shown below.

1. **Understand Equipment:** Begin by comprehensively understanding the equipment you want to monitor. Identify critical components, failure modes, and historical data related to maintenance and breakdowns. Gather information on the manufacturer's recommendations, maintenance manuals, and any available sensor or diagnostic data.

2. **Define Key Performance Indicators (KPIs):** Determine the performance metrics that will be used to assess the health and condition of the equipment. These KPIs could include variables like temperature, pressure, vibration, noise, or any other relevant parameters that can indicate potential issues.

3. **Select Monitoring Techniques:** Identify the appropriate monitoring techniques and technologies that will enable you to collect data on the defined KPIs. This could involve using sensors, data loggers, or advanced monitoring systems. Consider each technique's cost, feasibility, and effectiveness in relation to your specific equipment and operational requirements.

4. **Set Baseline and Thresholds:** Establish a baseline for normal equipment behavior based on historical data or expert

knowledge. Determine thresholds for each KPI that, when exceeded, indicate a potential problem. These thresholds can be derived from statistical analysis, equipment specifications, or established industry standards.

5. **Implement Data Collection**: Deploy the chosen monitoring techniques to collect real-time or periodic data from the equipment. This may involve installing sensors, connecting to existing data sources, or integrating with equipment control systems. Ensure the data collection process is automated, reliable, and synchronized with the equipment's operational cycles.

6. **Data Analysis and Modeling**: Analyze the collected data using various statistical and analytical methods. Look for patterns, trends, anomalies, or deviations from the baseline behavior. Apply predictive modeling techniques, such as machine learning algorithms or statistical models, to predict future equipment behavior and potential failures.

7. **Condition Monitoring Alerts**: Develop an alert system that triggers notifications or warnings when the monitored KPIs breach the established thresholds. This can involve email alerts, text messages, dashboard notifications, or integration with a Computerized Maintenance Management System (CMMS) or Enterprise Asset Management (EAM) system.

8. **Maintenance Action Planning**: When an alert is triggered, define appropriate maintenance actions based on the severity and urgency of the situation. This could include scheduling preventive maintenance tasks, conducting further diagnostic tests, ordering spare parts, or initiating repair procedures.

9. **Continuous Improvement**: Regularly review and refine your predictive maintenance plan based on feedback, analysis of historical data, and insights gained from the maintenance

actions taken. Continuously update and optimize the monitoring techniques, KPIs, thresholds, and maintenance strategies to improve the accuracy and effectiveness of the predictive maintenance program.

10. **Documentation and Training**: Document all aspects of your predictive maintenance plan, including equipment specifications, monitoring techniques, data analysis procedures, maintenance actions, and results. Provide comprehensive training to the maintenance staff, operators, and relevant personnel to ensure they understand the plan, can interpret the alerts, and execute the appropriate maintenance tasks.

Remember that developing an effective predictive maintenance plan is an iterative process. It requires ongoing data collection, analysis, and refinement to improve the accuracy of predictions and optimize maintenance strategies over time.

In conclusion, developing a predictive maintenance plan requires a strategic and structured approach. By clearly identifying the goals and objectives, thoroughly assessing the current capabilities and resources, and creating a detailed and flexible roadmap, organizations can pave the way for a successful and effective predictive maintenance program.

Chapter 9: Getting Started on a Predictive Maintenance Program

An organization seeking to adopt predictive maintenance as part of a life cycle sustainment plan should follow the following steps. From the organization's perspective, leadership will be focused on four things.

1. **Objective**: What do I want to achieve?
2. **Resources**: What assets can I devote?
3. **Time**: How much time do we have?
4. **Cost**: What can I afford?

Consider goals and objectives in terms of key performance parameters. I want to improve my asset availability and reduce downtime. I seek to increase efficiency and decrease costs. I want to prepare my workforce and customers for disruption from artificial intelligence (AI) and natural disasters, enrich the customer experience, and protect jobs through reskilling.

The next step is to assign a program champion or manager to plan and implement the program. Provide the program manager with team members who have the skill set to succeed. Identify and assign resources to the team, software, hardware, IoT sensors, communications, and funding.

Divide the program into phases; consider naming the phases minimal, now, then, and later. This allows resources (manpower, equipment, and funding) to be affordable, scalable, and acceptable to the company, workforce, and customers over time.

Affordability is always the most difficult. You may have to focus on what disruption costs by doing nothing. A percentage of the operating budget is usually devoted to the sustainment of the company's operations, product line, supply chain, and husbanding resources. This is perhaps the first place to look when searching for funds to measure affordability.

Developing a predictive maintenance strategy starts with an implementation plan. Realize that all AI-centered efforts require a plethora of data. The pile of data needed may seem daunting but approach it in small bites. Much of the data you will need is already being captured. The following steps will assist the program manager in collecting the appropriate and available data.

Five data types may require consideration for a predictive maintenance program, they are:

1. **Transactional Data:** This data describes your core business activities. It may include the data of your purchasing and selling activities, production activities data, delivery and shipping data, and data related to hiring and firing employees.
2. **Master Data:** It consists of key information that makes up the transactional data. Master data usually contains places, parties, & things. It is relatively constant. Transaction data is created at lightning speed; the master data is constant.
3. **Reference Data:** Reference data is a subset of master data. It is usually standardized data that is governed by specific codifications. Reference data is much less volatile than master data.
4. **Reporting Data:** It's an aggregated data compilation for analysis and reporting. This data consists of transactional, master, and reference data.
5. **Metadata:** It's data about data. Metadata provides context to your data sets. It includes information like the date of data creation, who created the data, its source, and any changes made. To organize metadata, it's a good practice to include fields for this in your database tables for both master and transactional data.

Start by gathering the available data and ascertaining if that is sufficient to initiate a PdM program. The following description can assist with this determination. See Figure 19.

1. **Transactional:** collect the organization's operational & maintenance data accessible and *__existing now__*.
2. **Master:** gather locations, people, and *__assets unique__* to your organization. There is already a lot of master data in your database.

3. **Reference:** many organizations already store and use this data from previous projects and ***general knowledge, i.e., wikis and*** maps.
4. **Reporting:** important objective data, ***dashboard stuff***, what the technician and customer are accustomed to, familiar to clients or industry standards. i.e., financials, work orders, asset control tags.
5. **Meta:** requirement; you will need the ***data dictionaries*** on all data info assembled from the customer and organization, i.e., units, type, and length.

The Predictive Maintenance Program utilizes a sixth data type known as predictive data.

Predictive Data: This is the data that feeds your predictive maintenance models. It usually includes sensor data from your machines (like temperature, pressure, vibration, etc.), but it might also include other data types (like weather data, production data, etc.). Predictive data needs to be organized to support efficient access and processing, often requiring a time-series database or a big data platform. See Figure 19.

Remember that data quality is key in predictive maintenance. You should have procedures in place to regularly check and clean your data. You should also have appropriate data governance measures to ensure data privacy and compliance with relevant regulations. Moreover, the structure and organization of your data should be flexible enough to accommodate changes in your assets, business, and maintenance strategies.

Besides these specific considerations, ensure that:

- Your data is reliable, consistent, and of good quality. Any data errors or inconsistencies can significantly affect your predictive maintenance model's accuracy.

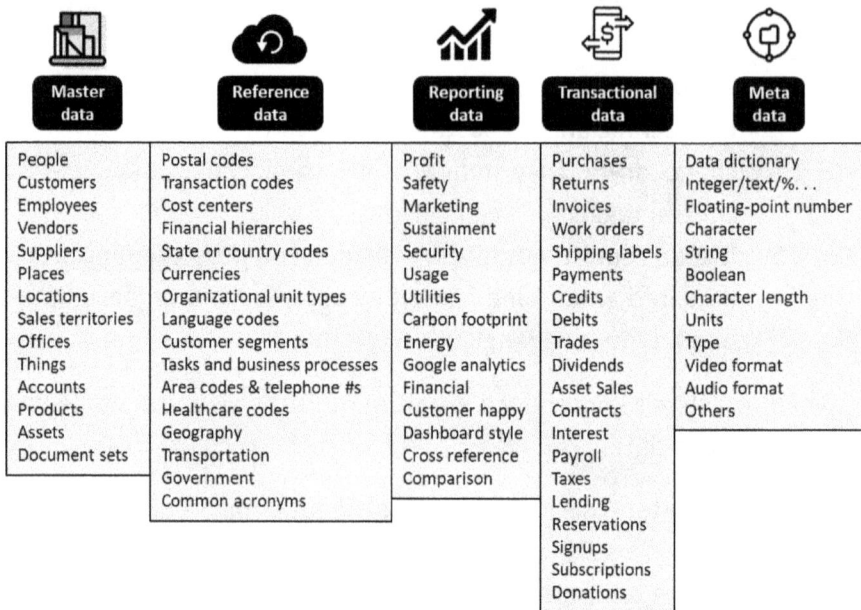

Master data	Reference data	Reporting data	Transactional data	Meta data
People	Postal codes	Profit	Purchases	Data dictionary
Customers	Transaction codes	Safety	Returns	Integer/text/%...
Employees	Cost centers	Marketing	Invoices	Floating-point number
Vendors	Financial hierarchies	Sustainment	Work orders	Character
Suppliers	State or country codes	Security	Shipping labels	String
Places	Currencies	Usage	Payments	Boolean
Locations	Organizational unit types	Utilities	Credits	Character length
Sales territories	Language codes	Carbon footprint	Debits	Units
Offices	Customer segments	Energy	Trades	Type
Things	Tasks and business processes	Google analytics	Dividends	Video format
Accounts	Area codes & telephone #s	Financial	Asset Sales	Audio format
Products	Healthcare codes	Customer happy	Contracts	Others
Assets	Geography	Dashboard style	Interest	
Document sets	Transportation	Cross reference	Payroll	
	Government	Comparison	Taxes	
	Common acronyms		Lending	
			Reservations	
			Signups	
			Subscriptions	
			Donations	

Figure 19: Examples of Data Elements for 5 Data Types

- Data security and privacy are upheld. Ensure you follow the necessary standards and legal requirements to protect your data.

- The data is accessible to all necessary parties, like data analysts, engineers, or machine learning models.

- Data architecture is scalable. As the amount of data collected grows, your architecture should be able to handle it.

- You maintain data lineage information, which is the data's lifecycle, including where it came from and how it has been changed over time. This is especially important for traceability and audit purposes.

Finally, consider implementing a data governance framework to manage your data's availability, usability, integrity, and security. This will help to maintain the quality and reliability of your predictive maintenance program.

Chapter 10: Execution and Scaling Predictive Maintenance Program

The execution phase is where the well-laid plans for predictive maintenance start to materialize. Beginning with a pilot project and gradually scaling to full-scale implementation ensures the plan's effectiveness and allows room for adjustments. The integration of predictive maintenance into current processes and systems forms a key aspect of this phase, ensuring seamless operation and optimization.

From Pilot to Full-Scale Implementation: Starting with a pilot project is a recommended approach in executing a predictive maintenance plan. Depending on the organization's size and complexity, the pilot could focus on a particular piece of equipment, a specific production line, or a given site. The main aim is to validate the predictive maintenance concept in a controlled environment and to learn and adapt before scaling up. The insights and lessons learned from the pilot project can guide the roll-out of the full-scale implementation, enabling adjustments to the plan, mitigating risks, and increasing the likelihood of success.

When transitioning to **Full-Scale Implementation**, it's essential to maintain the momentum generated during the pilot phase. The organization should leverage the successes and lessons from the pilot to gain stakeholders' buy-in and refine the processes and systems as they're scaled up. Effective communication, training, and support are essential during this transition to ensure that all staff understand the changes, see the benefits, and are equipped to contribute to the predictive maintenance program.

Implementing a predictive maintenance strategy is a complex process that involves careful planning, testing, and evaluation before a full-scale roll-out. Below are the steps from a pilot phase to full-scale implementation. Also, see Figure 20.

1. Identification and Planning: Before starting a pilot, organizations must identify the equipment or system on which predictive maintenance will be applied. Key performance indicators (KPIs) that will be used to evaluate the effectiveness of the predictive maintenance strategy must also be determined. In addition, the organization must define the scope of the pilot and its objectives.

2. Data Collection and Analysis: Historical data is gathered from the selected equipment or system and then used to identify patterns and trends. Sensors may need to be installed to collect real-time data. The data collected is then used to build predictive models and algorithms to forecast potential failures.

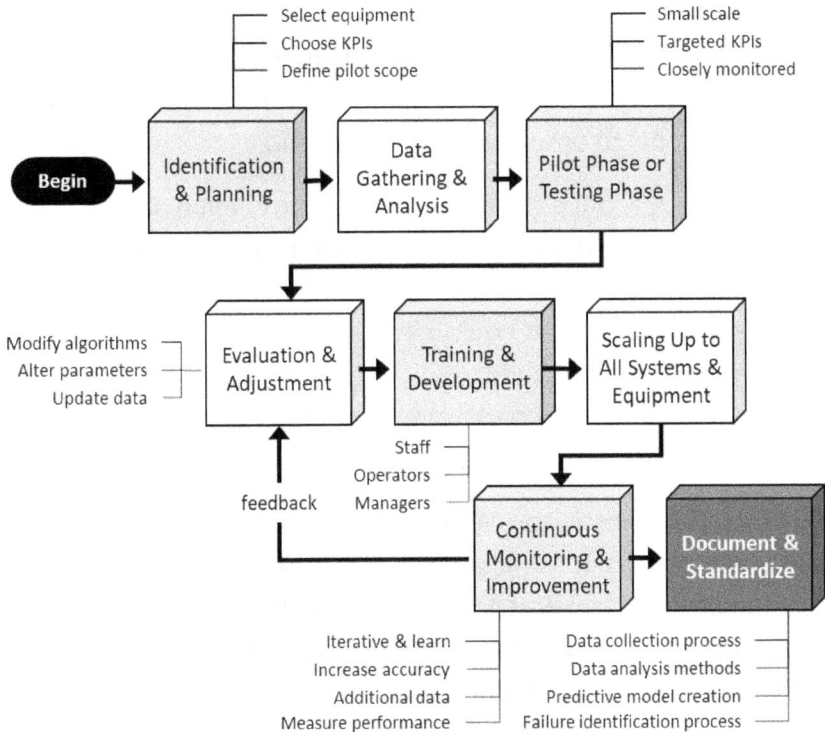

Figure 20: Pilot to Full-Scale Implementation Journey

3. Pilot Phase: This is the testing phase. The predictive maintenance strategy is applied on a small scale to the selected equipment or system. The strategy's performance is then closely monitored to assess its effectiveness based on the predefined KPIs.

4. Evaluation and Adjustment: After the pilot phase, the organization assesses the results to see if the predictive maintenance strategy achieved its objectives. The strategy is then refined based on the outcomes of the pilot phase. Adjustments might include modifying the predictive algorithms, altering the parameters used in the predictive model, or updating the data collection methods.

5. Training and Development: Before moving to full-scale implementation, it's vital to ensure that all stakeholders – including maintenance staff, operators, and managers – understand the predictive maintenance strategy and know how to use the system. Training sessions, workshops, and educational materials can be valuable tools to ensure a smooth transition.

6. Scaling Up: Once the strategy has been refined and all stakeholders have been trained, the organization can proceed to full-scale implementation. This involves applying the predictive maintenance strategy to all relevant equipment and systems in the organization. This might be done gradually, with the plan being expanded to additional equipment or systems over time, or it might be done all at once.

7. Continuous Monitoring and Improvement: Even after full-scale implementation, it's crucial to continue monitoring the performance of the predictive maintenance strategy. Regular evaluations can help identify any areas where the plan can be improved. As the system collects more data, the predictive models can be updated and refined to increase their accuracy further.

8. Documentation and Standardization: For full-scale implementation, it is essential to document the procedures, routines, and tasks in a standardized way. This includes documenting the data collection process, data analysis methods, predictive model creation, failure identification process, and maintenance task planning.

Remember that the whole process is iterative, and it's often necessary to return to earlier steps to make improvements and adjustments as needed. The key to a successful predictive maintenance strategy is continuous learning and improvement.

Integration with Current Processes and Systems is essential to ensure that predictive maintenance becomes a seamless part of the organization's operations. This involves integrating the predictive maintenance software with existing IT systems, aligning the predictive maintenance processes with current workflows, and training the staff to incorporate predictive maintenance tasks into their roles. The organization should aim for a holistic integration, where predictive maintenance is not a separate or additional task but an integral part of how the organization operates and maintains its equipment.

Continuous improvement should remain a core focus as the organization progresses on its predictive maintenance journey. The organization should regularly review the performance of the predictive maintenance program, seek feedback from staff, and use the insights gained to improve and evolve the program continuously. This iterative approach can help the organization to adapt to changes, tackle challenges, and maximize the benefits of predictive maintenance.

Chapter 11: Training and Building the Right Team

The success of a predictive maintenance program is not solely reliant on advanced technology and software; it equally depends on the people who operate, manage, and make decisions based on these tools. Building the right team with the necessary roles and skills and investing in training and continuous learning are crucial to implementing an effective predictive maintenance strategy.

Necessary Roles and Skills: A predictive maintenance team typically involves a mix of roles, each contributing unique skills and expertise. Key roles often include equipment operators who monitor the daily performance of machinery, maintenance engineers who carry out the actual maintenance tasks, data scientists who develop and manage the predictive models, and decision-makers who use the insights from the models to make strategic maintenance decisions.

Each of these roles requires specific skills. Equipment operators and maintenance engineers should have a deep understanding of the equipment and its operation while also being comfortable using predictive maintenance software to monitor equipment health and carry out maintenance tasks. Data scientists need strong skills in data analysis, machine learning, and possibly domain-specific knowledge, depending on the complexity of the equipment and the industry. Decision-makers need to understand the principles of predictive maintenance and the insights of predictive models to make informed, data-driven decisions.

Implementing a predictive maintenance program is a multifaceted task that requires an array of skills and roles within an organization. Key roles and their associated skills are listed below and shown in Figure 21.

1. **Leadership Team:** The C-suite or executive team is responsible for setting the strategic direction of the predictive maintenance program. They need to have strong decision-making skills, understand the business implications of predictive maintenance, and be able to lead the company through the change management process.

2. **Maintenance Managers/Engineers:** They have practical experience with maintaining machinery or systems and a deep understanding of the operations. They are crucial in defining what the predictive maintenance program should focus on. Skills include problem-solving, knowledge of maintenance processes and workflow, and experience with specific industry equipment.

3. **Data Scientists:** These individuals are tasked with developing the algorithms that predict when maintenance should be performed. Skills include machine learning, statistics, data modeling, and a strong understanding of the data being collected and how it relates to machine performance.

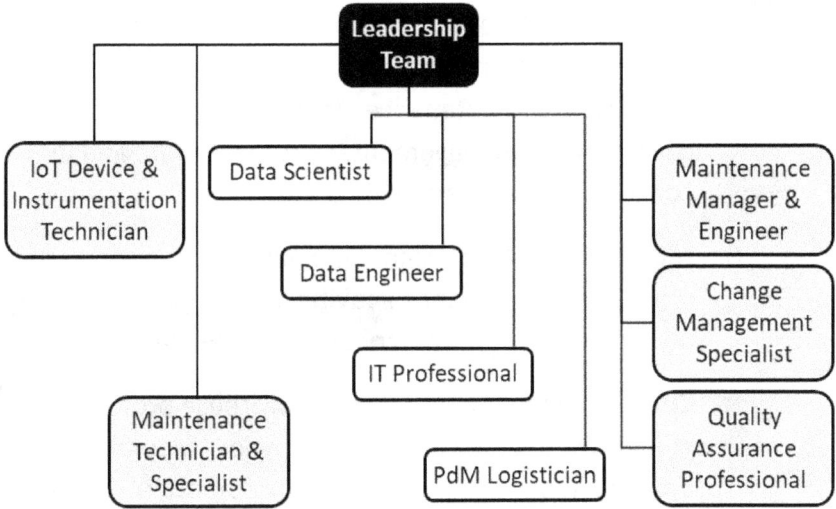

Figure 21: Suggested PdM Roles for Team Members

4. **Data Engineers:** They are responsible for managing the data pipeline and ensuring that data from the machines is appropriately collected, cleaned, and prepared for the data scientists. Skills include database management, data wrangling, and understanding of data privacy and security.

5. **IoT/Instrumentation Technicians:** These people install, maintain, and troubleshoot the sensors and other data collection devices on the machinery. Skills include an understanding of sensor technology, knowledge of the machines being instrumented, and the ability to work with data collection systems.

6. **IT Professionals:** They handle the integration of predictive maintenance software with existing IT systems. They also ensure system security and data privacy. Skills include IT infrastructure knowledge, cloud computing, and cybersecurity.

7. **Predictive Maintenance Logisticians:** Experts who will handle support equipment, parts ordering, consumables, packaging,

handling, and transportation for removal, replacement, and repair. They represent the supply chain management aspect of predictive maintenance. In addition, they perform the configuration management, documentation, and budgetary concerns for the leadership team.

8. **Maintenance Technicians/Specialists:** They carry out the predicted maintenance activities. They need to deeply understand the machinery or system being maintained.

9. **Change Management Specialists:** These professionals help guide the company and its employees through the changes brought about by implementing predictive maintenance. This includes training employees in new procedures, communicating the benefits of predictive maintenance, and managing resistance to change. Skills include communication, training, and understanding of organizational behavior.

10. **Quality Assurance Professionals:** They are responsible for monitoring the quality and effectiveness of the predictive maintenance program. They need to understand both the technical aspects of predictive maintenance as well as the operations of the machinery.

Implementing a predictive maintenance program involves more than just technical expertise; it also requires change management skills, leadership, and a strong understanding of the maintenance operations. By having a team with diverse skills, an organization can ensure the successful implementation of a predictive maintenance program. See Figure 22.

Data Analysis	Quality Control	Technical Skills
Predictive Modeling	Computer & IT Skills	
Machine Learning	Cybersecurity Vigilance	Project Management
Domain Knowledge	Database Management	
Problem-Solving	Supply Chain Mgmt	Data Capture Skills
Communication	Logistics & Sustainment	

Predictive Maintenance (PdM) Management Experience & Skillsets

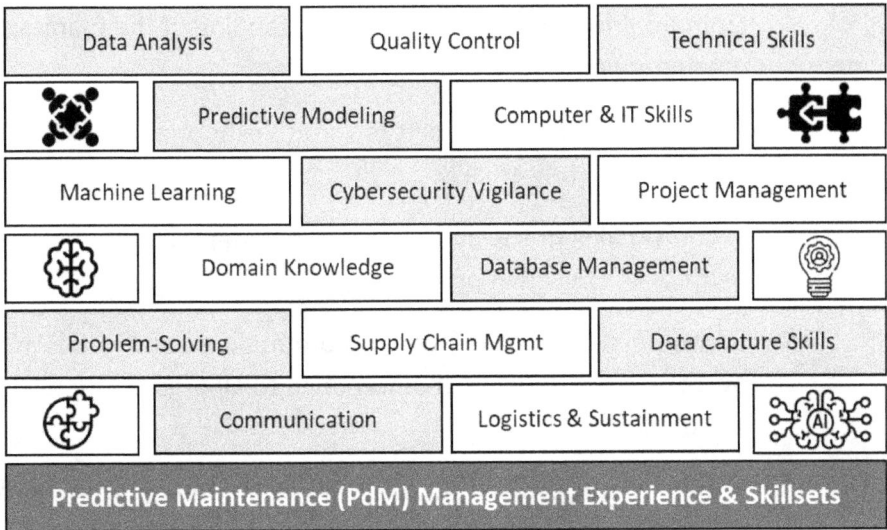

Figure 22: Suggested Skillsets & Expertise for PdM Team

Importance of Training and Continuous Learning: Given the technical nature of predictive maintenance and the advanced skills required, training is crucial to implementing a predictive maintenance program. This training should cover the basics of predictive maintenance, the use of predictive maintenance software, the interpretation of the data and insights, and the integration of predictive maintenance tasks into everyday roles and responsibilities.

Implementing a PdM program involves a multidisciplinary approach and requires various specific skill sets within the organization. A few interdisciplinary skillsets that should be considered when forming your PdM team are listed below.

1. **Data Analysis:** To interpret the data collected from machines.

2. **Predictive Modeling:** To predict when a machine will need maintenance.

3. **Machine Learning:** To build models that predict maintenance needs based on the collected data.

4. **Domain Knowledge:** A deep understanding of the maintained machinery or system is vital.

5. **Problem-Solving:** The ability to troubleshoot and solve problems as they arise.

6. **Technical Skills:** The ability to use and maintain the technologies used in predictive maintenance.

7. **Communication:** To effectively communicate the needs and outcomes of predictive maintenance to all organizational stakeholders.

8. **Project Management:** To oversee the implementation of the predictive maintenance program and ensure that it achieves its objectives.

Remember, successfully implementing a predictive maintenance program also requires cultural acceptance within the organization. It involves changing how maintenance is traditionally approached and hence needs buy-in from all levels of the organization.

Continuous learning is equally important in a field like predictive maintenance, which is continually evolving due to technological advancements and growing experience and knowledge in the field. Organizations should encourage and facilitate continuous learning through regular training sessions, workshops, conferences, online courses, and other learning opportunities.

In conclusion, building the right team and investing in training and continuous learning are key ingredients in the success of a predictive maintenance program. Organizations can effectively implement and optimize predictive maintenance and realize its many benefits by ensuring the team has the necessary roles and skills and by fostering a culture of continuous learning.

Part IV: Challenges, Solutions, and Future of Predictive Maintenance

Chapter 12: Overcoming Challenges in Predictive Maintenance

Embarking on a predictive maintenance journey comes with its share of challenges. These challenges often include data, cultural, technological, and organizational aspects. Understanding these potential obstacles and knowing how to address them is crucial to implement and optimize predictive maintenance successfully. See Figure 23.

Data, IT, AI Challenges Technology Challenges Organizational Challenges Cultural Challenges

Figure 23: Overcoming Challenges in Predictive Maintenance

Data Challenges: One of the first hurdles that organizations often face relates to data. This can include issues around data availability, quality, and volume. Sometimes, the necessary data may not be available or difficult to collect due to equipment or sensor limitations. Even when data is available, it may need to be of sufficient quality due to inaccurate sensor readings, missing data, or inconsistent data collection methods. These challenges can be addressed by investing in suitable sensors and data collection methods, implementing rigorous data cleaning processes, and using data augmentation techniques when needed.

Working with maintenance data can be crucial for many organizations but poses several challenges. Previous chapters of this book discuss the importance of data to perform analytic operations. These data challenges can sometimes be showstoppers. Best practices indicate that the earlier an organization addresses the data issues, the more efficient and cost-effective the data challenge becomes. Some of the most common ones are listed below. See Figure 24.

1. **Data Collection**: The first challenge is data collection. Often, data may be collected or tracked unevenly across an organization, especially in organizations that have a decentralized maintenance approach. This can lead to consistent data, which can make analysis easier.

2. **Data Quality**: Even when data is collected, it might not be accurate or complete. There can be inconsistencies due to human error, gaps in data collection, or obsolete data. Poor data quality makes it hard to make reliable decisions based on the data.

3. **Data Integration**: Integrating maintenance data from different sources and formats can be challenging. Some organizations use multiple maintenance systems for different operations, each storing data in a different way. Integrating these datasets can require significant time and effort.

4. **Data Standardization**: There is often a need for more standardization in maintenance data. Different parts of the organization might use different names or codes for the same piece of equipment or the same maintenance operation. This lack of standardization can make it challenging to analyze the data company-wide.

5. **Data Analysis**: Even when an organization has high-quality, integrated, standardized maintenance data, analyzing this data can be challenging. It requires the right tools and expertise to interpret the data and extract valuable insights.

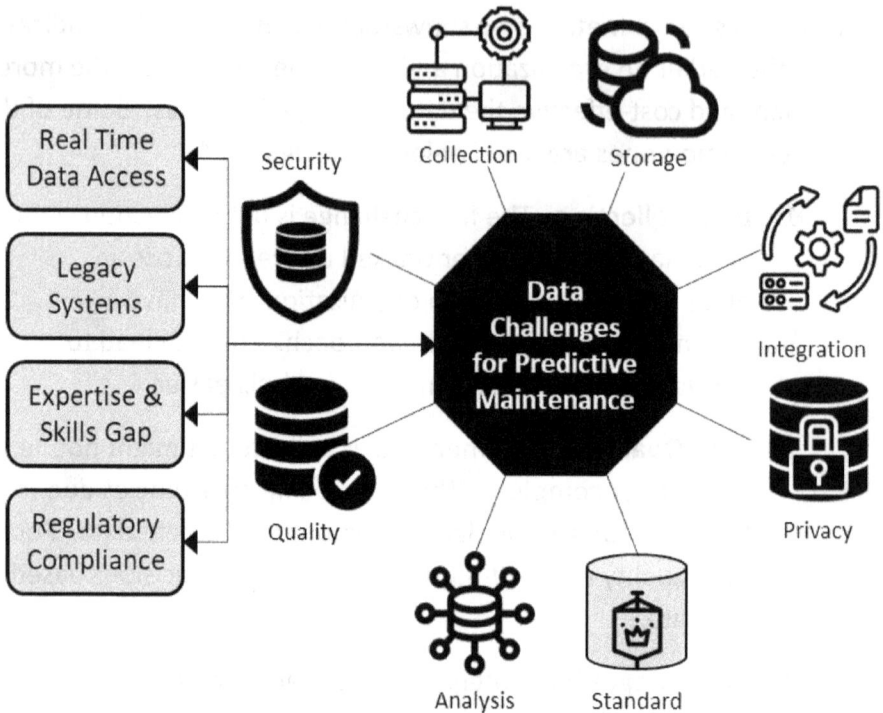

Figure 24: Examples of Data Challenges for PdM Implementation

6. **Data Security**: Maintenance data can be sensitive, particularly if it pertains to critical infrastructure. Ensuring the security of this data is a significant concern. It is always a good suggestion for your maintenance team to practice cyber hygiene to protect your data. Another good proposal is to back up your maintenance data often (every day).

7. **Data Storage**: As maintenance operations become increasingly digitized, the amount of data generated can be vast. Storing this data in a way that is accessible and easy to manage can be a challenge.

8. **Real-Time Data Access**: In many cases, having access to real-time maintenance data is crucial for making timely decisions.

However, making this possible can be technically challenging and resource intensive.

9. **Legacy Systems**: Many organizations still use legacy systems for their maintenance operations. These systems may not be designed to work with modern data analytics tools, making it difficult to extract and use maintenance data effectively.

10. **Data Privacy**: Depending on the nature of the maintenance data, privacy concerns may arise, especially if the data involves personally identifiable information. Organizations need to navigate these concerns while still effectively using their maintenance data.

11. **Expertise and Skills Gap:** Effective data management requires a combination of IT, data analytics, and domain knowledge. Many organizations need a skills gap in these areas, making it difficult to leverage their maintenance data effectively.

12. **Regulatory Compliance:** In specific industries, maintenance data must be recorded and maintained in accordance with regulatory requirements. Meeting these requirements while using the data for operational decision-making can be a significant challenge.

Addressing these challenges often requires a strategic approach to data management, along with the right tools and expertise.

Cultural Challenges: The shift from reactive or preventive maintenance to predictive maintenance represents a significant cultural change. This change can be met with resistance from employees who are used to traditional methods or skeptical about new technologies. Overcoming this requires clear communication about the benefits of predictive maintenance, effective change management practices, and ongoing training and support to help employees adapt to the new way of working.

Adopting a predictive maintenance program is about more than just technology. It also involves overcoming a range of cultural issues within an organization. Several critical cultural challenges are described below. See Figure 25.

1. **Resistance to Change**: This is a common issue in many organizations. People can be comfortable with how things currently work and may resist adopting new technologies or methods. It's essential to effectively communicate the benefits of predictive maintenance to all stakeholders and involve them in overcoming this resistance.

2. **Distrust in Technology**: Predictive maintenance relies on advanced technologies such as AI and machine learning. Some employees may be skeptical about the accuracy or reliability of these technologies, mainly if they've been used to relying on their intuition or experience. Building trust in these technologies is a crucial step.

3. **AI and Trend Analysis Skills Gap**: Predictive maintenance requires new skills, such as data analysis, working with AI, and interpreting predictive insights. There may be a lack of these skills within the organization, and some employees may feel threatened or undervalued if their current skills aren't relevant in the new system. Providing training and upskilling opportunities can help address this issue.

4. **Silos in Organization**: Predictive maintenance often requires cross-functional collaboration, as it involves areas like maintenance, operations, and IT. However, many organizations operate in silos, with each department focused on its own goals. Breaking down these silos and encouraging collaboration is essential for a successful predictive maintenance program.

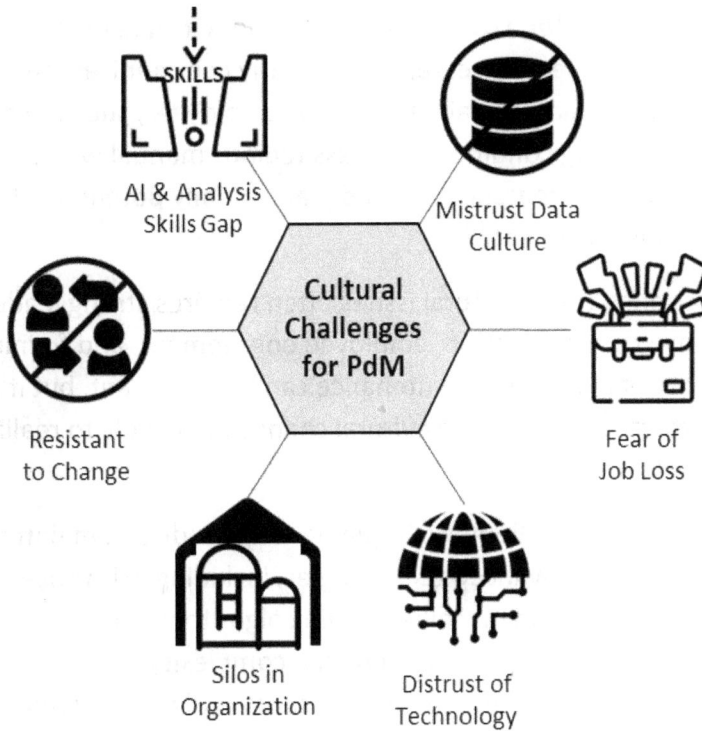

Figure 25: Cultural Challenges for PdM Implementation

5. **Short-term Thinking**: Predictive maintenance often requires significant upfront investment and may not deliver immediate results. Some organizations may focus on short-term profits and be reluctant to make this investment. Encouraging a long-term view can help overcome this issue.

6. **Data Culture**: Predictive maintenance relies on data, but not all organizations have a strong data culture. Some organizations may not be used to making decisions based on data, or there may be a lack of data literacy. Building a strong data culture and improving data literacy is key to adopting predictive maintenance.

7. **Fear of Job Loss**: Automation and AI can lead to worries about job losses. It's essential to clearly communicate the benefits of predictive maintenance for employees, such as safer working conditions and less tedious manual work, and to provide reassurances and plans for any potential job displacement.

Overcoming these cultural issues often requires strong leadership, effective communication, employee engagement, and training. The benefits of predictive maintenance can be significant, but it's important to manage the cultural change effectively to realize these benefits.

Technological Challenges: Implementing predictive maintenance involves using advanced technologies, including IoT devices, data analytics tools, and machine learning algorithms. These technologies can present challenges regarding the complexity of implementation, the need for new skills, and data security and privacy issues.

Organizations can overcome these challenges by seeking expert help, investing in in-house training to develop the necessary skills, and implementing robust data security measures. Consider the following technological challenges when adopting a predictive maintenance program. See Figure 26.

1. **Algorithm Development**: Predictive maintenance requires sophisticated machine learning or statistical models to predict future equipment failures. Developing these models involves data science and machine learning expertise, which may be lacking in many organizations.

2. **Real-time Processing**: For predictive maintenance to be effective, it must often operate in real-time. This can be technically challenging, especially for companies with large, complex operations.

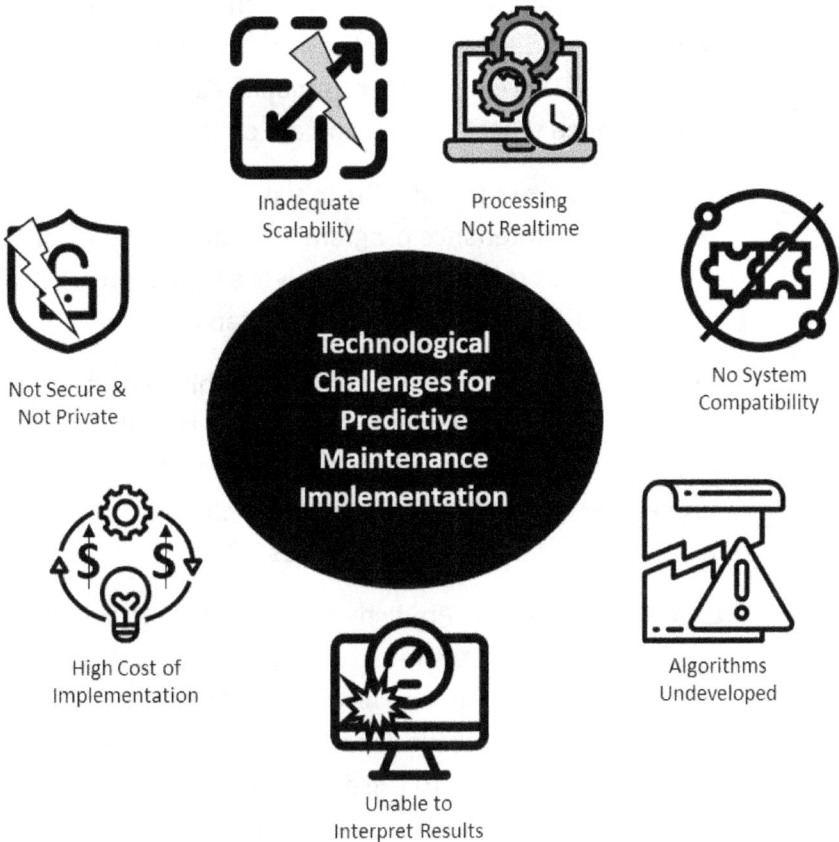

Figure 26: Technological Challenges for PdM Implementation

3. **Security and Privacy**: With the increasing adoption of IoT devices, the risk of cyber-attacks also increases. Ensuring the security of data, mainly when it's transmitted across networks, is a significant challenge.

4. **System Compatibility**: In many instances, existing infrastructure may not support implementing a predictive maintenance program. Integrating the new system with legacy systems may be difficult, and upgrading these systems can be costly and time-consuming.

5. **Interpretability of Results**: Machine learning algorithms can sometimes behave as "black boxes," making it difficult for humans to understand how they arrived at a particular prediction. This lack of interpretability can make it hard for maintenance teams to trust and act on the predictions.

6. **Cost of Implementation**: The initial cost of implementing a predictive maintenance program, including the cost of sensors, data storage, processing, and analytics capabilities, as well as training staff to use the new system, can be high.

7. **Scalability**: Lastly, as the number of monitored assets grows, the predictive maintenance system must be able to scale accordingly. Technology needs to support more extensive data processing and analysis, which can be challenging for rapidly expanding or large-scale businesses.

Organizational Challenges: Predictive maintenance is not just a technological initiative but a strategic business change that affects various parts of the organization. It requires alignment and collaboration between different departments, clear roles and responsibilities, and changes to existing processes and workflows. Overcoming these challenges often requires strong leadership, clear communication, and continuous willingness to learn and adapt. A few of the more exciting challenges are provided below. See Figure 27.

1. **Leadership Support**: With the proper support and understanding from leadership, implementing predictive maintenance can be smooth. Leaders need to understand the benefits and ROI of predictive maintenance to justify the required investment.

2. **Change Management**: Implementing predictive maintenance represents a significant change in how maintenance is performed. Organizations must manage this change

effectively, ensuring that all stakeholders understand their new roles and responsibilities.

3. **Budget Allocation**: Predictive maintenance requires an initial investment in sensors, data storage and processing infrastructure, and analytics tools. It might be difficult to secure this funding, mainly if the organization is not entirely convinced of the value of predictive maintenance.

4. **Lack of Defined Processes**: To successfully implement predictive maintenance, organizations must have transparent processes for data collection, analysis, and action. These processes in place are necessary for the predictive maintenance initiative to deliver the desired results.

5. **Misalignment between Departments**: There can be a disconnect between different departments, such as operations, maintenance, IT, and data science teams. Each team needs to work together to ensure the success of the predictive maintenance program.

6. **Regulatory Compliance**: In some industries, predictive maintenance must be implemented in compliance with regulations. This can complicate the implementation process and necessitate additional oversight.

7. **Reliability of Predictions**: If the predictive maintenance system provides false alarms or misses impending failures, it can lead to distrust in the system and a return to older maintenance methods.

8. **Data Privacy and Security Concerns**: Given that predictive maintenance relies heavily on data, data privacy and security concerns can pose challenges. Ensuring that all data used complies with privacy regulations and is secure from cyber threats is a crucial challenge.

Case Studies on Successful Problem-Solving: Examining case studies from organizations that successfully implement predictive maintenance can provide valuable insights and practical solutions to these challenges. For example, a manufacturing company might share how they overcame data challenges by implementing new sensor technology and robust data-cleaning processes. Or a utility company might share how they addressed cultural challenges by running a successful change management program. These case studies can offer practical tips, inspire new ideas, and reassure that the challenges can be overcome.

In conclusion, while predictive maintenance comes with challenges, it can be effectively addressed with the right strategies, continuous learning, and a resilient mindset. Overcoming these challenges can pave the way for successful predictive maintenance implementation and drive broader improvements in the organization's operations and culture. See Figure 27 for examples of organizational challenges when implementing predictive maintenance.

Figure 27: Organizational Challenges for Predictive Maintenance Implementation

1. **General Electric (GE)**: GE has long been a leader in applying predictive maintenance in its operations, specifically in the realm of its wind turbine business. They deployed digital

twins and predictive analytics to model the performance of individual turbines. As a result, they were able to anticipate maintenance needs and act before issues could cause downtime. GE mitigated the challenge of algorithm development by investing heavily in its data science and analytics teams, which had the expertise to develop predictive models.

2. **Thyssenkrupp Elevators**: One of the world's leading elevator companies implemented a predictive maintenance solution for their elevators called MAX. By equipping elevators with IoT sensors and using Microsoft Azure's machine learning capabilities, they can predict maintenance needs and reduce elevator downtime. Thyssenkrupp addressed the challenges of data integration and quality by collaborating with a technology partner who could provide the necessary expertise and tools.

3. **Royal Dutch Shell**: Shell launched a predictive maintenance project in its offshore oil drilling operations. It applied machine learning algorithms to sensor data from thousands of machines, predicting potential equipment failures up to weeks in advance. Shell addressed the challenges of real-time data processing and storage by investing in robust cloud infrastructure and forming partnerships with tech companies specialized in advanced analytics and machine learning.

4. **Siemens**: Siemens implemented a predictive maintenance system for their train fleet known as Railigent. Railigent uses sensor data and machine learning to predict when components might fail and recommend maintenance before that happens. Siemens tackled the problem of system compatibility by building a cloud-based system that could interface with their existing hardware, reducing the need for extensive upgrades.

5. **Delta Airlines**: Delta Airlines has been using predictive maintenance to minimize the downtime of its aircraft. They equipped planes with sensors to monitor various parts and systems, and the data gathered helps predict potential failures. They mitigated the challenge of data security and privacy by employing stringent data governance practices and partnering with security firms.

In all these cases, the businesses mitigated challenges through strategic partnerships, investment in necessary technology and expertise, and a commitment from leadership to change management and adoption of new methods. They also focused on data governance and security to ensure compliance with regulations and protect critical operational data.

Research provides some other examples listed below. The company's names aren't shared for privacy reasons. These representative examples of small business types originate from different industry sectors. They have successfully adopted predictive maintenance, improved performance, and reduced costs.

1. **Small Manufacturing Business:** A small-scale manufacturing company producing automotive parts integrated with IoT sensors in their production machinery. This allowed them to collect data on machine performance and identify anomalies indicative of potential issues. The implementation of predictive maintenance not only helped them reduce machinery downtime but also enabled them to increase their production rate, improve the quality of parts, and reduce costs associated with sudden machine failures. See Figure 28.

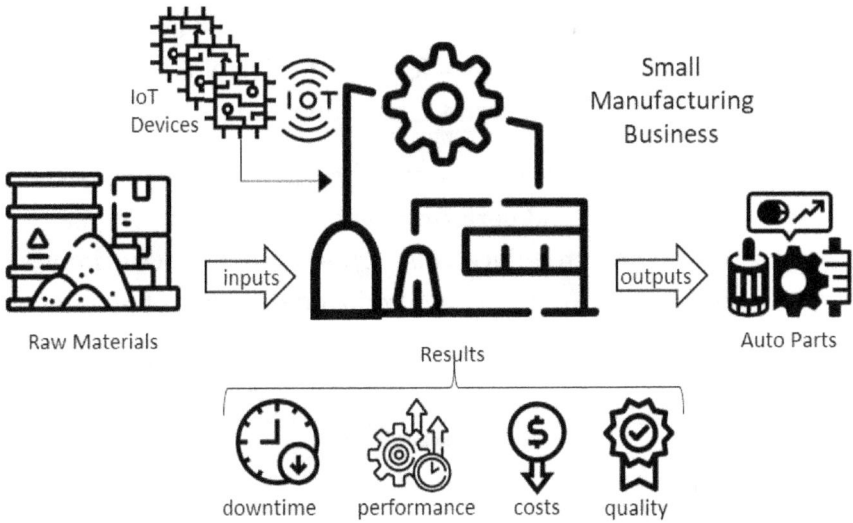

Figure 28: Small Manufacturing Business PdM Example

2. **Agricultural Business:** A small farming operation implemented predictive maintenance on their farm machinery like tractors, combines, and irrigation systems. Using IoT and AI-based predictive analysis, they managed to prevent equipment failures during critical farming periods. This increased overall productivity. See Figure 29.

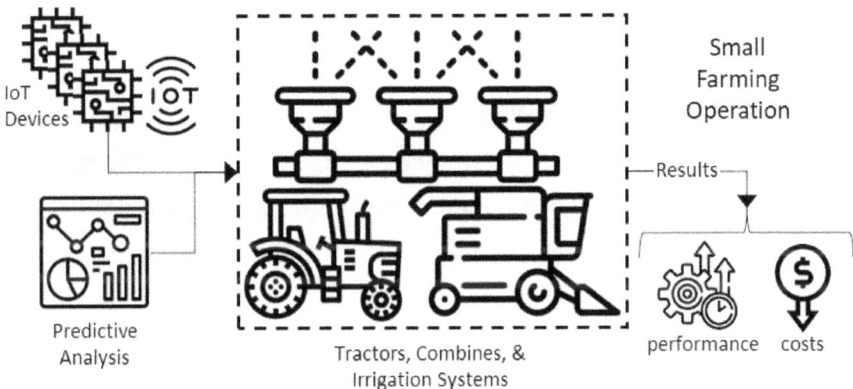

Figure 29: Small Agricultural Business PdM Example

3. **Specialty Coffee Shop:** A small local coffee chain adopted predictive maintenance for its critical equipment, including espresso machines and refrigeration units.

   Using data from IoT devices, they could predict potential breakdowns and perform maintenance during off-peak hours. This approach reduced disruption to their service, improved customer experience, and reduced the costs of emergency repairs. See Figure 30.

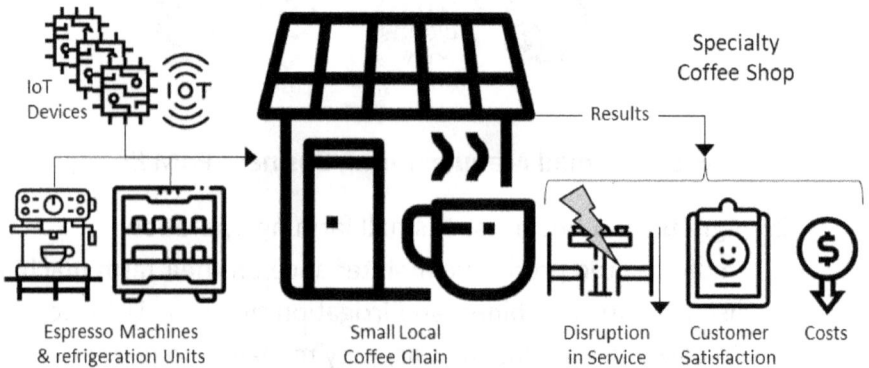

Figure 30: Specialty Coffee Shop PdM Example

4. **Local Delivery Service:** A small local delivery service implemented predictive maintenance on their fleet of delivery vehicles.

   Using onboard diagnostics and telematics data, they were able to anticipate maintenance needs, minimizing downtime and reducing costs associated with emergency repairs and part replacements. See Figure 31.

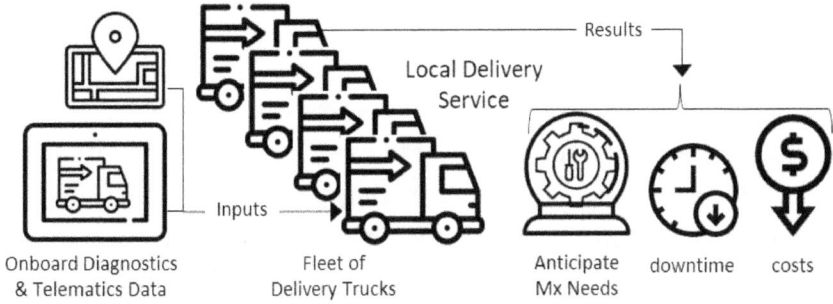

Figure 31: Local Delivery Service PdM Example

In each of these examples, the businesses mitigated challenges such as budget constraints, lack of in-house technical expertise, and cultural resistance through a combination of strategies. These included incremental implementation, working with expert vendors, extensive training, and communication to demonstrate the value and benefits of predictive maintenance to all stakeholders.

Chapter 13: Measuring Success and ROI

Measuring success and ROI (Return on Investment) is crucial for any strategic initiative like predictive maintenance. It helps justify the investment, guides continuous improvement, and provides valuable feedback to all stakeholders. Key performance indicators (KPIs) and metrics play a central role in quantifying the effectiveness and benefits of predictive maintenance.

It helps to re-emphasize the importance of this section with the famous quote by Peter Drucker, "If you can't measure it, you can't manage it." One more time, Lord Kelvin's quote, "If you can't measure it, you can't improve it." It's impossible to know whether your predictive maintenance efforts are successful unless success is defined and tracked. With an established metric for success, progress can be quantified, and adjustments can be made to produce the desired outcome. See Figure 32.

Measuring the Improvement in Predictive Maintenance Effectiveness involves assessing the performance of the predictive maintenance program and comparing it with previous or alternative maintenance approaches. Some methods commonly used to measure predictive maintenance improvement are listed below. Two things are critical now: (1) choose the key important parameters to your company. Next, (2) record a baseline – a starting point that can be used as a benchmark.

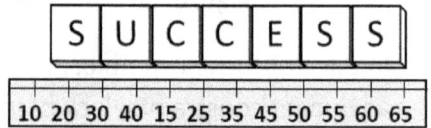

| S | U | C | C | E | S | S |

10 20 30 40 15 25 35 45 50 55 60 65

Figure 32: Measure Success

1. **Equipment uptime and availability:** One of the primary goals of predictive maintenance is to minimize unplanned equipment downtime. You can assess the program's effectiveness by tracking and comparing the uptime and availability of equipment before and after implementing predictive maintenance strategies. Decreased downtime and increased availability indicate improved maintenance practices.

2. **Mean time between failures (MTBF):** MTBF is a metric that measures the average time between equipment failures. By monitoring MTBF over time, you can assess whether predictive maintenance has increased the interval between failures. A higher MTBF indicates improved reliability and maintenance effectiveness.

3. **Mean time to repair (MTTR):** MTTR measures the average time required to repair equipment after a failure. Monitoring MTTR can help evaluate the efficiency and effectiveness of predictive maintenance activities. A decrease in MTTR

suggests that the predictive maintenance program has improved the speed and efficiency of repairs.

4. **Reduction in reactive maintenance:** Reactive maintenance refers to addressing failures or issues as they occur. One of the critical objectives of predictive maintenance is to shift from reactive maintenance to proactive maintenance. By measuring the reduction in reactive maintenance work orders or emergency repairs, you can assess the success of the predictive maintenance program in preventing unexpected failures.

5. **Cost savings:** Predictive maintenance can lead to cost savings by minimizing unexpected failures, optimizing maintenance schedules, and reducing unnecessary maintenance activities. By comparing the maintenance costs before and after implementing predictive maintenance, you can evaluate the financial impact and cost-effectiveness of the program.

6. **Overall equipment effectiveness (OEE):** OEE is a comprehensive metric that considers factors such as equipment availability, performance efficiency, and product quality. By measuring OEE before and after implementing predictive maintenance, you can evaluate the impact on overall equipment performance. An increase in OEE indicates improvement in maintenance effectiveness.

7. **Maintenance backlog:** Tracking the backlog of maintenance work orders or pending repairs can provide insights into the efficiency and effectiveness of the maintenance program. Reducing the maintenance backlog suggests that predictive maintenance has improved the prioritization and execution of maintenance tasks.

8. **User satisfaction and feedback:** Gathering feedback from maintenance personnel, equipment operators, and other

stakeholders involved in the maintenance process can provide qualitative insights into the perceived improvement of predictive maintenance. Conducting surveys or interviews to assess user satisfaction can help identify areas of improvement and further optimization.

It's important to note that measuring predictive maintenance improvement requires careful data collection, analysis, and comparison with baseline or historical data. The specific metrics and methods will depend on the goals, available data, and industry-specific requirements.

KPIs and Metrics for Predictive Maintenance: To measure the success of a predictive maintenance program, organizations should define and track a set of relevant KPIs and metrics. These can vary depending on the specific goals and context of the organization, but standard metrics often include:

- **Reduction in Downtime:** One of the main goals of predictive maintenance is to reduce unplanned downtime by anticipating equipment failures before they happen. Comparing downtime before and after the implementation of the predictive maintenance program can help measure its effectiveness.

- **Increased Equipment Lifespan:** Predictive maintenance can help identify minor issues before they become major problems, leading to longer equipment lifespans. Monitor the average lifespan of your equipment before and after the program to measure improvements.

- **Maintenance Costs:** Predictive maintenance should reduce planned and unplanned maintenance costs by ensuring maintenance is only performed when necessary. Compare the costs associated with maintenance before and after the program.

- **Return on Investment (ROI):** Calculate the ROI by dividing the net benefit (benefits - costs) by the investment cost. Benefits could include reduced maintenance costs, downtime, and extended equipment life.

- **Mean Time Between Failures (MTBF):** The average time between system breakdowns. A successful predictive maintenance program should increase the MTBF.

- **Predictive Accuracy:** Measure how accurate your predictive maintenance model predictions are. This might include measuring the number of true positives (correctly predicted failures), false positives (predicted failures that didn't happen), true negatives (correctly predicted non-failures), and false negatives (failures that weren't predicted).

- **Efficiency Metrics:** These could include production throughput, energy consumption, and product quality, which can all be affected by equipment health. Improvements in these metrics can indicate effective predictive maintenance.

- **Scheduled Maintenance Compliance:** Track the percentage of planned maintenance tasks completed on schedule. An increase in this metric may indicate the successful implementation of predictive maintenance.

Remember, to have a fair evaluation, your measures need to be compared to a baseline. This baseline can be the performance measures before implementing the predictive maintenance program or industry standards. Also, it might take some time to see improvements, as predictive maintenance programs can take time to implement fully and start showing results.

Quantifying Savings and Improvement: To calculate predictive maintenance's Return on Investment (ROI), organizations need to quantify the savings and improvements achieved through the program. This involves comparing the current performance metrics with the baseline metrics from before the predictive maintenance program was implemented. The difference represents the improvements or savings attributable to predictive maintenance.

For example, if the equipment uptime has increased from 90% to 95% since implementing predictive maintenance, then this 5% increase represents an improvement in production capacity. If the maintenance cost has been reduced from $100,000 per year to $80,000, this $20,000 saving represents a direct financial benefit.

In addition to these quantifiable benefits, predictive maintenance can bring fewer tangible benefits, such as improved staff morale, better decision-making, and increased customer satisfaction. While these benefits may be more complex to quantify, they should not be overlooked in evaluating the predictive maintenance program's success and ROI. See Figure 33 for a baseline to model output comparison process.

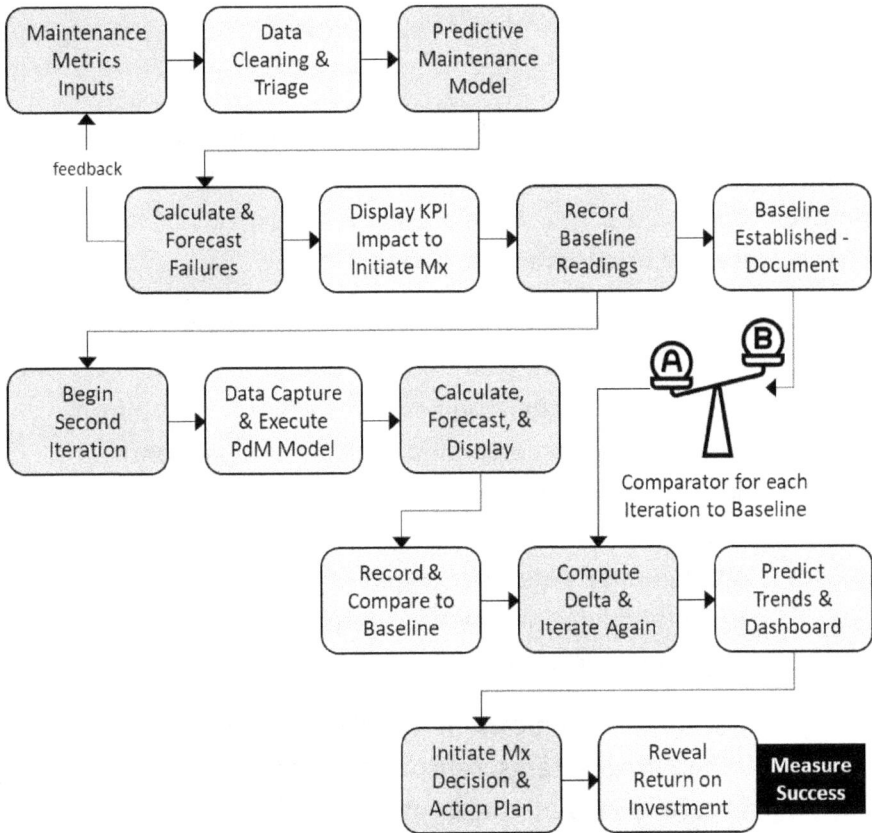

Figure 33: Establish Baseline for Comparison to Measure Success

In conclusion, measuring success and ROI is crucial to implementing a predictive maintenance program. By defining relevant KPIs and metrics and quantifying the savings and improvements achieved, organizations can validate the effectiveness and benefits of predictive maintenance, justify the investment, and guide the continuous improvement of the program.

Chapter 14: Predictive Maintenance and Cybersecurity

Cybersecurity plays a crucial role in predictive maintenance to ensure the integrity, confidentiality, and availability of data and systems involved in the process. Here are some key aspects of cybersecurity in the context of predictive maintenance:

1. **Data Protection:** Predictive maintenance relies on collecting and analyzing large amounts of data from sensors, monitoring systems, and other sources. Protecting this data from unauthorized access, tampering, or loss is essential. Implement strong access controls, encryption techniques, and secure data storage practices to safeguard the data throughout its lifecycle. Figure 34 depicts data generators - Cyber and PdM (augmented with AI and ML). The result is improved performance.

2. **Secure Communication:** The data transmission between monitoring systems, sensors, and analytics platforms must be secure to prevent interception or manipulation. To encrypt data in transit, utilize secure communication protocols, such as HTTPS or MQTT with TLS/SSL. Protect the integrity of the data by implementing mechanisms to detect and prevent unauthorized modifications during transmission.

3. **Device Security**: Ensure the devices' security in the predictive maintenance ecosystem. This includes sensors, data loggers, gateways, and other hardware components. Apply security best practices, such as regularly updating firmware and software patches, implementing robust authentication mechanisms, and disabling unnecessary services or ports.

4. **The integrity of Predictive Models**: The models used in predictive maintenance are often based on machine learning, which requires clean, accurate data for training and operation.

If a cyber-attack contaminates this data, the resulting models may give false predictions, leading to unplanned downtime or unnecessary maintenance work.

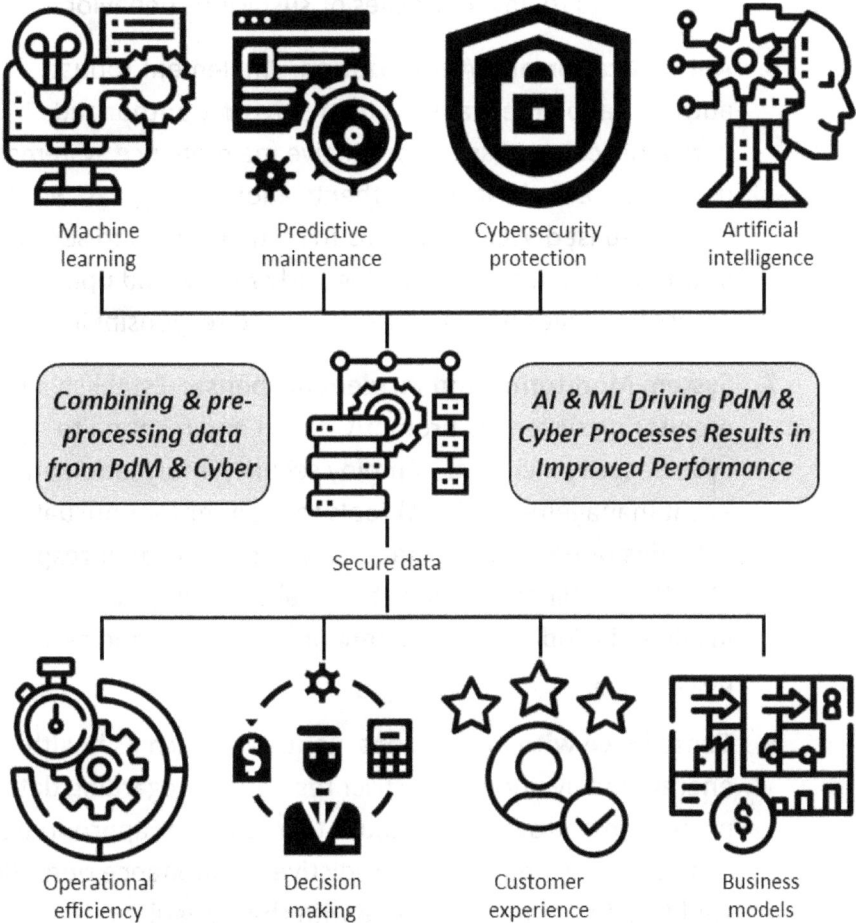

Figure 34: AI & ML-Driven Data Processes Improve Performance in Several Areas

5. **Network Security**: The infrastructure supporting predictive maintenance activities should be secure to prevent

unauthorized access or potential attacks. Implement firewalls, intrusion detection and prevention systems, and network segmentation to protect the infrastructure from external threats. Regularly monitor and log network activity to detect and respond to any anomalies or suspicious behavior.

6. **Authentication and Authorization**: Implement robust authentication mechanisms to ensure only authorized personnel can access the predictive maintenance systems and data. Utilize multi-factor authentication, strong passwords, and role-based access controls to restrict access to sensitive information and functions. Regularly review and update user access privileges based on job roles and responsibilities.

7. **System Monitoring and Incident Response**: Establish a robust monitoring system to promptly detect and respond to cybersecurity incidents. Employ security information and event management (SIEM) tools to monitor for unusual activities or potential threats. Develop an incident response plan that outlines the steps to be taken during a cybersecurity incident, including containment, investigation, and recovery procedures.

8. **Compliance with Regulations**: Industries such as healthcare, energy, and aviation have strict regulations regarding data protection and systems. Failure to implement appropriate cybersecurity measures in predictive maintenance operations could lead to non-compliance with these regulations, potentially resulting in fines or other penalties.

9. **Confidence in the System**: Users must trust the predictive maintenance system to be effective. Cybersecurity breaches can undermine this trust, making it harder for the system to deliver its benefits.

10. **Avoidance of Catastrophic Failures**: In industries where equipment failure can lead to catastrophic outcomes (e.g., nuclear power plants, aviation, etc.), cybersecurity is even more critical. Cyber-attacks could result in manipulated data, leading to inaccurate predictions and potential disasters.

11. **Vendor and Supply Chain Management**: If you are working with third-party vendors or suppliers for predictive maintenance solutions, ensure they adhere to robust cybersecurity practices. Conduct due diligence to evaluate their security posture, including data protection measures, security certifications, and vulnerability management processes.

12. **Employee Training and Awareness**: Train employees and stakeholders on cybersecurity best practices in the predictive maintenance process. Create awareness about potential threats, social engineering techniques, and the importance of following secure procedures. Encourage prompt reporting of any suspicious activities or incidents.

By prioritizing cybersecurity in predictive maintenance, organizations can mitigate the risk of data breaches, system compromises, and potential disruptions to critical operations.

It helps ensure the reliability and trustworthiness of the predictive maintenance system and protects sensitive information throughout the process.

Figure 35 provides a wire diagram combining predictive maintenance and cyber technologies to protect data and forecast failures.

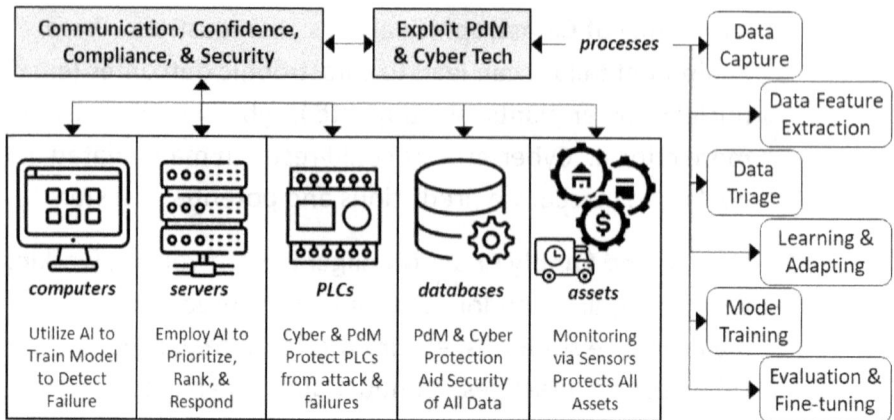

computers	servers	PLCs	databases	assets
Utilize AI to Train Model to Detect Failure	Employ AI to Prioritize, Rank, & Respond	Cyber & PdM Protect PLCs from attack & failures	PdM & Cyber Protection Aid Security of All Data	Monitoring via Sensors Protects All Assets

Figure 35: PdM & Cyber Processes Provide Data Protection & Failure Forecasting to Improve Performance

There is another aspect linking cybersecurity and predictive maintenance. Predictive maintenance and cybersecurity can intersect to predict failures and cyber-attacks in several ways. Combining these two disciplines can create a secure and efficient IT infrastructure. Several strategies are listed below where these two technologies intersect for more significant benefit.

1. **Predicting Hardware and Software Failures:** Predictive maintenance can help anticipate hardware and software failures by monitoring system health. It uses machine learning to predict potential system crashes based on patterns and trends, such as repeated errors, abnormal resource consumption, etc. When paired with cybersecurity, these insights can be used to identify if these irregularities are natural or due to malicious activities, like malware or hacker infiltration.

2. **Detecting Unusual Patterns:** Predictive analytics can be used to understand standard behavioral patterns within a network or system. This could be based on network traffic, user behavior, or system performance. Any deviations from this

typical pattern could indicate a potential issue. For example, if there is a sudden surge in network traffic or abnormal system behavior, it could indicate a cyber-attack. Figure 35 illustrates how the data collection and preprocessing of PdM, and cyber data align with each other.

3. **Predicting Cyber Attacks:** By analyzing historical data on cyber-attacks, predictive models can be developed to predict future attacks. These models could identify trends and patterns in the type of attacks, the most vulnerable areas, etc. This can help develop robust defense mechanisms and prioritize the areas that require immediate attention.

4. **Predictive Analysis for System Health**: Just as predictive maintenance uses data from machinery to forecast potential failures, similar principles can be applied to IT systems. Monitoring key performance indicators, such as system load, memory usage, CPU usage, network traffic patterns, and other factors, can help anticipate potential failures. Machine learning algorithms can be trained to identify patterns in these metrics that predict system crashes, allowing IT teams to intervene before a failure occurs.

5. **Anomaly Detection**: By establishing what is 'normal' behavior in a system, detecting anomalies that may signal an impending cyber-attack is possible. For example, an unusually high volume of network traffic could suggest a DDoS attack, while an unexpected increase in failed login attempts might point towards a brute force attack. Machine learning can be used to refine the sensitivity of this anomaly detection, reducing the number of false positives while still quickly identifying real threats. See Figure 36.

6. **Risk Assessment**:
Predictive analytics can assess the risk associated with different systems and data sets. Factors like the age of the system, past attack history, the sensitivity of the data it holds, and its importance to the organization can all be used to predict where attacks are most likely to occur. This can help organizations prioritize their cybersecurity efforts, first protecting the most vulnerable and valuable parts of their infrastructure.

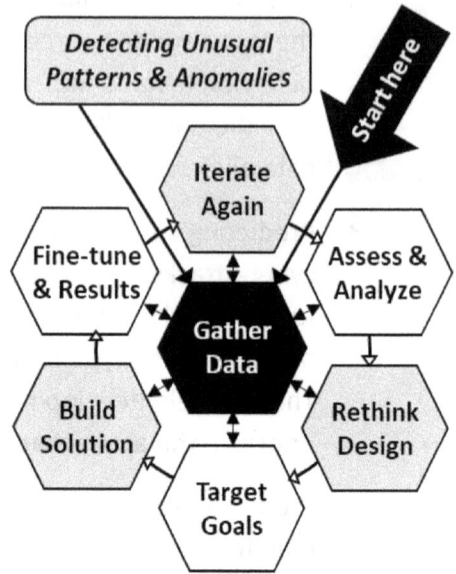

Figure 36: Combining Cyber & PdM Processes Improves Analysis & Fault Detection

7. **User Behavior Analysis**: User behavior analysis can be considered part of the predictive analytics model. It focuses on monitoring and analyzing the behavior of users within an organization's network. Any significant deviations from normal behavior might indicate insider threats or accounts that have been compromised.

8. **Threat Intelligence**: This involves gathering and analyzing information about current and emerging threats. Cybersecurity teams can use this intelligence to prepare for known attacks and update their predictive models with the latest data, making them more accurate.

9. **Simulation and Testing**: Cybersecurity teams can use predictive models to simulate different types of attacks, testing the resilience of their systems. This can help them to predict how their systems would react in the event of an actual attack and to identify any potential weaknesses that need to be addressed.

10. **Combining Cybersecurity and Operational Data**: One of the innovative ways of leveraging predictive maintenance for cybersecurity is to combine operational data (data about the system's performance, environmental conditions, etc.) with cybersecurity data (logins, network traffic, etc.). This combination can provide a holistic view of the system's operation, potentially uncovering correlations that might be missed when looking at the datasets independently. See Figure 37 for a data flow diagram supporting the power of analyzing PdM and Cyber data combined.

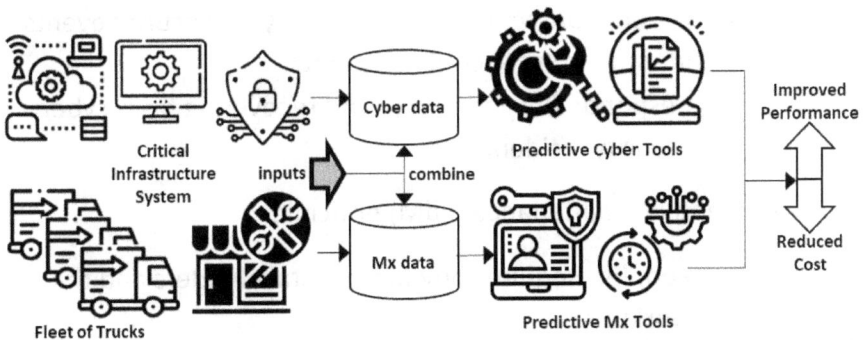

Figure 37: Combining PdM & Cyber Data is More Powerful

While these strategies can significantly reduce the risk and impact of cyber-attacks, it's important to note that no system can ever be entirely secure. Therefore, a robust cybersecurity strategy should also include measures to limit the damage of successful attacks and recover quickly when they occur.

Therefore, cybersecurity must be an integral part of any predictive maintenance strategy to ensure the system can operate effectively and safely. As stated previously, this includes secure data encryption, strong authentication, regular security audits, and ongoing staff training in cybersecurity best practices.

This discussion intends to highlight the new evolving capabilities of predictive maintenance (PdM) and the principles that can be applied to predictive cybersecurity. Integrating these methodologies provides several advantages and reaps the rewards of data collection from both disciplines.

The terms and methodology for failure and data breach forecasting are very similar. Note the two depictions below.

The cyber security principles:

- Govern: Identifying and managing security risks.
- Protect: Implementing controls to reduce security risks.
- Detect: Detect and understand cyber security events to identify cyber security incidents.
- Respond: Responding to and recovering from cyber security incidents.

The preventive maintenance (PdM) principles:

- People: Requires buy-in from stakeholders and the workforce.
- Data: The key element to tie history, present, and future together.
- Processes & Technology: The algorithms & IT to utilize, analyze, and formulate future vision.
- Balanced strategy: PM is a part of an overall maintenance program to reduce costs and improve performance.

In conclusion, PdM uses condition-monitoring tools and techniques and asset information to track real-time and historical equipment performance so you can anticipate failure before it happens. The challenge is to integrate these principles into a blended methodology.

A predictive approach that analyzes information from multiple sources enables organizations to identify and prevent cyberattacks before they happen. Predictive threat intelligence tools provide insights so security leaders can devote more resources to protecting their company's most vulnerable targets. Predictive maintenance tools give an understanding of equipment conditions, sense the historical and present status, and predict when a subsystem will fail.

Predictive analytics uses analyses that make predictions and projections about future events or trends to identify risks and better inform security protocols or defenses. Prediction involves forecasts about future events and trends that have not yet occurred.

Predictive modeling is a commonly used statistical technique to predict future behavior. Predictive modeling solutions are a form of data-mining technology that analyzes historical and current data and generates a model to help predict future outcomes. Technological advancements have given us extraordinary problems concerning information security and failure forecasting, but they also have given us the means to combat them.

Chapter 15: The Future of Predictive Maintenance

The journey of predictive maintenance is far from static. As an area underpinned by rapidly evolving technology, its landscape is ever-changing. Staying informed about emerging trends and technologies and building a predictive maintenance strategy that is adaptable and scalable are essential to future-proof your operations and maintain competitiveness.

Emerging Trends and Technologies: Several key trends shape the future of predictive maintenance. Machine learning and artificial intelligence continue to evolve and mature, enabling more sophisticated and accurate predictive models. These technologies are becoming more accessible and user-friendly, which may result in broader adoption and more varied applications in predictive maintenance.

Another significant trend is the growth of digital twin technology, where a digital replica of physical assets helps optimize maintenance strategies and predict potential issues before they occur. Additionally, edge computing is becoming increasingly relevant in predictive maintenance, enabling data analysis directly at the point of data collection, thus reducing data transmission time and allowing real-time analytics and actions.

Sustainment Planning: The artificial intelligence features used in predictive maintenance are critical to sustainment planning. Sustainment planning refers to the processes involved in maintaining and extending the useful life of equipment, systems, or infrastructure. With the rise of AI and machine learning technologies, there are several ways AI can be employed to improve and optimize sustainment planning within the context of predictive maintenance and failure forecasting.

1. **Predictive Maintenance:** As we mentioned earlier, PdM and AI can be used to predict when a device or system might fail. Machine learning algorithms can analyze vast amounts of sensor data to identify patterns or anomalies that indicate a potential future failure. This allows for timely maintenance, reducing downtime and costs.

2. **Resource Optimization:** PdM and AI algorithms can be used to optimize available resources. For instance, these technologies can help schedule maintenance activities at the

optimal time to minimize disruption or assist in managing spare parts inventory to ensure that the necessary parts are available when needed.

3. **Supply Chain Management:** PdM and AI can be used to optimize supply chains, ensuring that necessary materials and parts are available when needed. The supply system can better plan component availability by forecasting which parts will fail. This can involve predicting future demand for parts, optimizing the logistics of moving parts from one place to another, and identifying potential issues in the supply chain that could disrupt the availability of necessary parts.

4. **Cybersecurity Protection:** Prognostic analytics determine when and where attacks/failures may occur. The tech world has been searching for ways to keep sensitive data out of the hands of our adversaries. Until now, cybersecurity preventative maintenance has mainly been an after-the-fact fight, patching holes only noticed after a hacker attacks an organization. Then, mitigate the event by indicating where it hurts or adopt a repair and replace failed equipment strategy after it happens. This is too late to prevent the damage or loss of performance that transpires. Integrating cybersecurity, AI, ML, and predictive maintenance goes a long way in protecting data, forecasting failures, increasing availability, and preventing or mitigating cyber-attacks.

5. **Life Cycle Management:** PdM and AI/ML can help to model and predict the entire lifecycle of equipment or systems, from acquisition to disposal. This can provide valuable insights into the total cost of ownership, helping to inform decisions about when to replace equipment or make significant upgrades.

6. **Fault Detection and Diagnosis:** PdM employing AI and ML can not only predict failures but also help diagnose the cause of

failures when they occur. Machine learning algorithms can analyze data from various sources to identify the most likely cause of failure, which can help speed up repair times.

7. **Simulation and Training:** PdM utilizing AI & ML can create realistic simulations of various scenarios, which can be used for training purposes or to test different strategies. For instance, these technologies could simulate the impact of a new maintenance strategy on overall system performance.

8. **Risk Assessment and Management:** PdM using AI and ML algorithms can be used to predict and evaluate risks associated with different sustainment strategies. This involves assessing the likelihood of different types of failures, the potential impact of these failures, and the effectiveness of different strategies for mitigating these risks. See Figure 38 for a depiction of risk assessment in the future.

These are just a few examples. The potential applications of PdM combined with AI and ML in sustainment planning are vast and continue to grow as technology evolves. The key is to understand the specific challenges and needs of your sustainment planning processes and identify how AI can best address these.

Preparing for the Future: Adaptable and Scalable Predictive Maintenance: To stay ahead in this fast-evolving field, organizations should strive to make their predictive maintenance programs adaptable and scalable. Adaptable programs can incorporate new technologies and methods as they emerge, continually enhancing their capabilities and effectiveness. This requires an open and agile approach to technology, a strong focus on learning and innovation, and a willingness to experiment and take calculated risks.

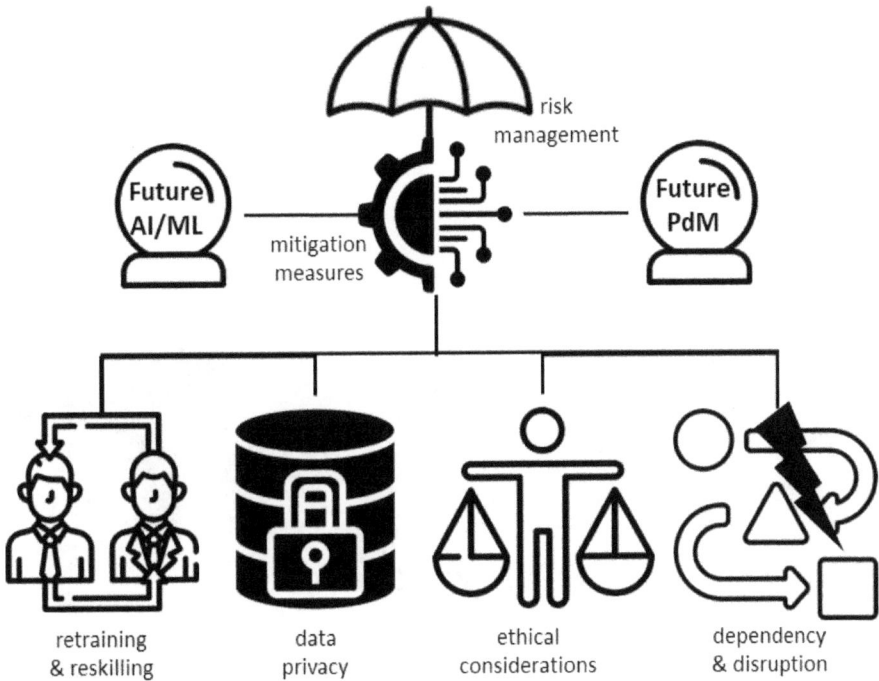

Figure 38: Future PdM Utilizing AI & ML to Improve Key Issues for Businesses

Scalable predictive maintenance programs can grow with the organization, handling larger volumes of data, more complex equipment, and more diverse operational scenarios. This requires robust and flexible IT infrastructure, efficient and scalable data management practices, and a strategic approach to growth that balances short-term gains with long-term sustainability.

In conclusion, the future of predictive maintenance is promising and exciting. By staying informed about emerging trends and technologies and building adaptable and scalable predictive maintenance programs, organizations can leverage the full potential of predictive maintenance and maintain a competitive edge in their industry.

Part V: Appendices

Appendix A: Glossary of Predictive Maintenance Terms

1. **Anomaly Detection**: Identifying data points, events, or observations that deviate from an established norm or pattern. Predictive maintenance is used to detect unusual equipment behavior that could indicate a potential issue.

2. **Artificial Intelligence (AI)**: The capability of a machine to imitate intelligent human behavior. In predictive maintenance, AI is used to analyze data and predict equipment failures.

3. **Data Augmentation**: The process of increasing the amount and diversity of data, often by creating synthetic data or introducing variations in existing data. This can improve the performance of predictive models in predictive maintenance.

4. **Data Cleaning**: The process of identifying and correcting or removing errors in data. This ensures that the data used in predictive maintenance is accurate and reliable.

5. **Descriptive Analysis**: The process of analyzing historical data to understand what has happened in the past. In predictive maintenance, this could involve analyzing past equipment failures to identify patterns or trends.

6. **Diagnostic Analysis**: The process of diagnosing the cause of a specific outcome or event. In predictive maintenance, this could involve diagnosing the cause of an equipment failure.

7. **Predictive Maintenance (PdM)**: It refers to the use of data-driven, proactive maintenance methods that analyze an equipment's condition to predict when it might fail. This allows maintenance to be planned before failure occurs.

8. **Condition Monitoring**: The process of monitoring a parameter of condition in machinery (vibration, temperature, etc.) to identify a significant change that is indicative of a developing fault.

9. **Failure Prediction**: It refers to the use of data, algorithms, and machine learning to predict when an equipment failure might occur based on patterns and trends.
10. **Machine Learning**: A type of artificial intelligence (AI) that allows software applications to become more accurate in predicting outcomes without being explicitly programmed. It is pivotal in predictive maintenance for identifying patterns and trends in machine behavior.
11. **Data Analytics**: The process of examining, cleaning, transforming, and modeling data with the goal of discovering useful information, forming conclusions, and supporting decision-making.
12. **Internet of Things (IoT)**: The network of physical devices, vehicles, buildings, and other items embedded with electronics, software, sensors, and network connectivity that enables these objects to collect and exchange data.
13. **Real-Time Monitoring**: The live monitoring of physical and digital processes to gather data and report on performance and issues instantly.
14. **Fault Detection**: The process in which the fault or defects in a system are detected. In predictive maintenance, this typically refers to finding defects or issues in machinery or equipment.
15. **Prognostics**: The part of the predictive maintenance process where, after data has been gathered and analyzed, the future state of a component or system is predicted.
16. **Predictive Model**: A statistical model or machine learning algorithm that is used to predict future behavior based on historical data.
17. **Anomaly Detection**: The identification of rare items, events, or observations which raise suspicions by differing significantly from much of the data.
18. **Digital Twin**: A virtual model of a process, product, or service. This pairing of the virtual and physical worlds allows analysis of

data and monitoring of systems to prevent problems before they occur.

19. **Health Monitoring**: In the context of predictive maintenance, health monitoring often refers to the continuous or periodic evaluation of a machine or system's status and performance to diagnose potential issues or predict future failures.

20. **Remaining Useful Life (RUL)**: The length of time a machine or component is expected to perform without failure given its current condition and performance.

21. **Asset**: Any item, thing or entity that has potential or actual value to an organization. This can refer to machines, systems, devices, etc.

22. **Operational Downtime**: The period a system is not available or not operating to its full potential due to maintenance, failures or outages.

23. **Reliability Centered Maintenance (RCM)**: A corporate-level maintenance strategy that is implemented to optimize the maintenance program of a company or facility. The result of an RCM program is the implementation of a specific maintenance strategy on each of the assets of the facility.

24. **Failure Modes and Effects Analysis (FMEA)**: A systematic method for evaluating a process to identify where and how it might fail and to assess the relative impact of different failures, in order to identify the parts of the process that are most in need of change.

25. **Prescriptive Maintenance**: Also known as RxM, this type of maintenance involves using historical data and real-time data from machines to predict and mitigate failures. Prescriptive maintenance can recommend actions that can be taken to prevent failures and to maximize the lifespan of machines.

26. **Artificial Neural Networks (ANN)**: A subset of machine learning, they are computing systems vaguely inspired by the biological neural networks that constitute animal brains. Such systems learn

(progressively improve performance) to do tasks by considering examples.

27. **Deep Learning**: A subfield of machine learning where neural networks — algorithms inspired by the human brain — learn from large amounts of data. While a neural network with a single layer can still make approximate predictions, additional hidden layers can help optimize the results.

28. **Supervised Learning**: A type of machine learning where the model is provided with labeled training data. The model learns from this data, and then applies what it has learned to new, unseen data.

29. **Unsupervised Learning**: Unlike supervised learning, in unsupervised learning the model is not provided with labeled data. It must find structure and relationships in the data on its own.

30. **Root Cause Analysis**: The process of discovering the root causes of faults or problems to identify an issue so that steps can be taken to manage those causes.

31. **Regression Analysis**: A set of statistical processes for estimating the relationships among variables. It includes many techniques for modeling and analyzing several variables.

32. **Data Mining**: The practice of examining large databases to generate new information. In predictive maintenance, data mining can be used to identify trends and patterns that might not be immediately apparent.

33. **Fault Diagnosis**: The process of tracing and correcting a fault in a machinery or system. In predictive maintenance, it's the act of identifying the specific failure after it's been detected.

34. **Edge Computing**: A distributed computing paradigm that brings computation and data storage closer to the location where it is needed, to improve response times and save bandwidth.

35. **Historian Database**: A type of time-series database designed to efficiently collect, store, and provide access to the large amounts

of time-stamped data that's typically generated in industrial processes and utility applications.

36. **Data Integration:** The process of combining data from different sources into a unified view. It allows for more effective analysis of data.

37. **Time-Series Data:** A sequence of data points indexed in time order. It is a common form of data in predictive maintenance, where sensor readings are recorded over time.

38. **Time-Series Analysis:** Methods for analyzing time series data to extract meaningful statistics and other characteristics. It's often used in predictive maintenance to identify patterns or trends.

39. **Feature Extraction:** The process of transforming raw data into features that can be used to develop predictive models. In predictive maintenance, this often involves turning sensor data into meaningful indicators of machine health.

40. **Ensemble Learning:** A machine learning concept in which multiple models are trained to solve the same problem and combined to get better results. It's often used in predictive maintenance to create more robust and accurate models.

41. **Reinforcement Learning:** A type of machine learning where an agent learns to make decisions by taking actions in an environment to maximize some notion of cumulative reward.

42. **Fault Tree Analysis (FTA):** A top-down, deductive failure analysis in which an undesired state of a system is analyzed using Boolean logic to combine a series of lower-level events.

43. **Cloud Computing:** The delivery of computing services—including servers, storage, databases, networking, software, analytics, and intelligence—over the Internet ("the cloud") to offer faster innovation, flexible resources, and economies of scale.

44. **Sensor Fusion:** The combination of sensory data or data derived from disparate sources such that the resulting information has less uncertainty than would be possible when these sources were used individually.

45. **Bayesian Networks**: A type of probabilistic graphical model that uses Bayesian inference for probability computations. Bayesian networks aim to model conditional dependence, and therefore causation, by representing conditional dependence by edges in a directed graph.
46. **Confidence Interval**: A range of values, derived from a statistical model, that is likely to contain the value of an estimated parameter.
47. **Data Normalization**: The process of structuring a relational database in accordance with a series of so-called normal forms to reduce data redundancy and improve data integrity.
48. **Precision and Recall**: These are evaluation metrics based on an information retrieval perspective of the confusion matrix for binary classification problems. Precision answers the question "What proportion of positive identifications was actually correct?", whereas recall answers "What proportion of actual positives was identified correctly?"
49. **Signal Processing**: The analysis, interpretation, and manipulation of signals. Signals of interest can include sound, images, time-series data collected from sensors, and any other quantities that change or vary over time.
50. **End of Life (EOL)**: In the context of manufacturing and product lifecycles, end of life (EOL) is the final stages of a product's existence. Predicting the EOL of a machine or component is a key part of predictive maintenance.

Appendix B: Resources and Further Reading

1. **Books**:

   - **"Predictive Maintenance in Dynamic Systems: Advanced Methods, Decision Support Tools and Real-World Applications"** by Riccardo Accorsi and Riccardo Manzini. This book provides an in-depth look into predictive maintenance, with a focus on real-world applications.

   - **"Maintenance, Replacement, and Reliability: Theory and Applications"** by Andrew K.S. Jardine and Albert H.C. Tsang. This comprehensive guide focuses on the theory of predictive maintenance along with practical applications.

   - **"Complete Guide to Preventive and Predictive Maintenance (Volume 1) Second Edition** by Joel Levitt. This book shares the best practices, mistakes, victories, and essential steps for success which the author has gleaned from working with countless organizations.

   - **"Predictive Maintenance IOT (Advanced Analytics WikiBooks Book 2)"** by Martin Sykora. This book discusses the Industrial Internet of Things (IIoT) where IoT connected sensors can collect real-time data of use to manufacturers and producers.

   - **"Predictive Maintenance in Smart Factories: Architectures, Methodologies, and Use-cases (Information Fusion and Data Science)"** by Tania Cerquitelli. This book presents the outcome of a project addressing the design and development of a plug-n-play end-to-end cloud architecture and enabling predictive maintenance of industrial equipment.

- **"An Introduction to Predictive Maintenance (Plant Engineering) 2nd Edition"** by R. Keith Mobley. This book introduces Predictive Maintenance that helps the industrial plant, processes, maintenance and reliability managers and engineers to develop and implement a comprehensive maintenance management program.

- **"Industrial Internet of Things: Cybermanufacturing Systems"** by Sabina Jeschke, Christian Brecher, Houbing Song, and Danda B. Rawat: This book provides a deeper understanding of the industrial internet of things, including predictive maintenance, and its applications in various industries.

- **"Introduction to Machine Learning with Python"** by Andreas C. Müller and Sarah Guido: A great resource for understanding the fundamentals of machine learning, a key component of predictive maintenance.

- **"Predictive Maintenance with MATLAB"** by MathWorks. This eBook helps you get started with predictive maintenance algorithm development with MATLAB® by explaining the terminology and providing access to examples, tutorials, and trial software.

2. **Websites:**

- **The Predictive Maintenance Group:** An online community offering resources, discussions, and insights into the latest trends and best practices in predictive maintenance.

- **The International Society of Automation (ISA)**: The ISA provides resources and training in various areas of automation, including predictive maintenance.

- **OpenAI Blog**: A collection of articles and papers about the latest developments in artificial intelligence and machine learning, which are integral to predictive maintenance.

- **"An Executive's Guide to Machine Learning"** published on McKinsey & Company's website: This resource offers a clear and concise explanation of machine learning for non-technical readers.

- **"Predictive Maintenance: What is Predictive Maintenance?"** published on IBM's website: This article offers an easy-to-understand explanation of predictive maintenance and its benefits.

- **OpenAI's GPT-3**: AI's language model offers a multitude of use cases and one of them is predictive maintenance. A good place to start exploring more advanced uses of AI in predictive maintenance.

3. **Research Papers and Journals**:

   - **"A Review of Predictive Maintenance Techniques and Their Applications"** (Published in IEEE Access)

   - **"Predictive Maintenance in Industry 4.0: State-of-the-art Methods for Maintenance Prognostics"** (Published in The Journal of Manufacturing Systems)

   - **"A Review of Predictive Maintenance Algorithms and Their Applications"** published in Journal of Mechanical Science and Technology: This review paper gives a comprehensive overview of different predictive maintenance algorithms.

- **"Predictive Maintenance Modeling for Diverse Industry Applications"** published in Journal of Industrial Information Integration: This paper highlights the role of predictive maintenance in various industrial applications.

- **"Predictive Maintenance and Sensitivity Analysis for Equipment with Multiple Quality States"** by Xiao Wang. This paper discusses the predictive maintenance (PM) problem of a single equipment system. It is assumed that the equipment has deteriorating quality states as it operates, resulting in multiple yield levels represented as system observation states.

- **"The Journey to Predictive Maintenance"** by Vic Ramdass, Assistant Secretary of Defense – Sustainment. Predictive Maintenance is Condition-Based Maintenance Plus Condition-Based Maintenance is maintenance based on the evidence of need (EON).

- **"Model of a Performance Measurement System for Maintenance Management"** by José Contreras. This paper presents a model for indicators based on various hierarchical levels and different functions and processes taking place in a maintenance department.

- **"Predictive Maintenance Adoption: Overcoming Common Data Quality Issues"** by Justin Gagne. When it comes to manufacturing, downtime is the major productivity killer Predictive maintenance that are used by businesses that maximize their throughput can save as much as 40% when compared to inferior maintenance strategies.

- **"9 Common HVAC Problems You Can Avoid With Regular Maintenance"** by Enoch. If you perform regular maintenance on your HVAC system and schedule regular checkups with a trained technician, many common HVAC issues can be avoided.

4. **Online Courses:**

- **Predictive Maintenance and Analytics Course on Coursera:** This course provides an overview of predictive maintenance, including methods for data analysis and modeling.

- **Professional Certificate in IoT and Predictive Maintenance on edX:** This professional certificate program covers the key concepts of IoT and predictive maintenance.

- **"Introduction to Predictive Maintenance"** on Coursera: This course provides a beginner-level understanding of predictive maintenance, its benefits, and its implementation.

- **"Machine Learning for Predictive Maintenance"** on edX: This course dives deeper into the use of machine learning in predictive maintenance.

- **"Industrial IoT on Google Cloud Platform"** on Coursera: This course offers a broader view of the Industrial Internet of Things, including predictive maintenance.

- **"Maintenance KPIs and maintenance metrics – Turning numbers into action,"** on Fixsoftware.com:

Key performance indicators (KPIs) measure the performance of a person, department, project, or company over time, and are effective at achieving organizational objectives.

5. **Conferences and Events:**

   - **The Predictive Maintenance Conference**: An annual conference bringing together industry leaders, practitioners, and researchers to discuss the latest trends and developments in predictive maintenance.

   - **The Industrial Internet of Things (IIoT) Summit**: This event often features discussions and presentations on predictive maintenance as part of its focus on the broader topic of industrial IoT.

These resources offer further exploration of the field of predictive maintenance, from academic research to practical applications, online communities, and training opportunities. They can provide deeper insights, additional perspectives, and up-to-date information on the latest trends and developments.

Appendix C: HVAC Example

Title: Utilizing Predictive Maintenance for Commercial HVAC Systems

Introduction: Predictive maintenance has emerged as a valuable strategy for optimizing the performance and reliability of commercial HVAC (Heating, Ventilation, and Air Conditioning) systems. By employing advanced technologies and data analysis techniques, predictive maintenance can help identify and address potential failures proactively. This example explores the typical failures in commercial HVAC systems, their life expectancy, and the key metrics used to measure maintenance performance, including Mean Time to Repair (MTTR) and Mean Time Between Failures (MTBF).

Typical Failures in Commercial HVAC Systems: Commercial HVAC systems are complex and consist of numerous components that can experience various failures. Some of the typical failures include:

1. **Compressor Failures:** Compressors are vital components of HVAC systems that can experience issues such as motor failures, refrigerant leaks, or electrical faults.

2. **Fan and Motor Failures:** Fans and motors can degrade over time due to wear and tear, leading to reduced airflow, increased energy consumption, and potential system breakdowns.

3. **Sensor and Control System Failures:** Malfunctioning sensors or control systems can result in inaccurate temperature or humidity regulation, leading to discomfort for occupants and reduced energy efficiency.

4. **Refrigerant Leaks:** Leaks in the refrigerant circuit can cause a decline in cooling or heating capacity and may even lead to system breakdowns.

Life Expectancy of HVAC Systems: The life expectancy of a commercial HVAC system depends on various factors, including the quality of equipment, maintenance practices, and operating conditions. On average, a well-maintained HVAC system can last between 15 to 25 years. However, certain components, such as compressors, may have a shorter lifespan of around 10 to 15 years.

MTTR and MTBF: Mean Time to Repair (MTTR) and Mean Time Between Failures (MTBF) are essential metrics used to evaluate maintenance performance in commercial HVAC systems.

1. **Mean Time to Repair (MTTR):** MTTR refers to the average time taken to repair a failed component or restore a system to its normal operating condition. By monitoring and minimizing MTTR, organizations can reduce downtime and ensure optimal system performance.

2. **Mean Time Between Failures (MTBF):** MTBF represents the average time between consecutive failures in a system. A higher MTBF indicates better reliability and longer intervals between failures. Predictive maintenance techniques aim to increase MTBF by detecting and addressing potential issues before they cause a system breakdown.

Utilizing Predictive Maintenance for Commercial HVAC Systems: Predictive maintenance utilizes advanced technologies such as IoT (Internet of Things) sensors, data analytics, and machine learning algorithms to collect and analyze real-time data from HVAC systems. By monitoring key performance indicators, such as temperature, pressure, and energy consumption, predictive maintenance systems can identify deviations from normal operating conditions, predict failures, and recommend maintenance actions. This proactive approach helps organizations schedule maintenance activities efficiently, reduce downtime, optimize energy consumption, and extend the lifespan of HVAC equipment.

Conclusion: Predictive maintenance has proven to be a valuable approach for commercial HVAC systems. By leveraging advanced technologies and data analysis techniques, organizations can detect and address potential failures before they occur, thereby improving system reliability, reducing downtime, and optimizing energy efficiency. With a focus on minimizing MTTR, increasing MTBF, and utilizing predictive maintenance strategies, businesses can ensure the smooth operation of their commercial HVAC systems, providing comfort for occupants and cost savings in the long run. See Figure 39.

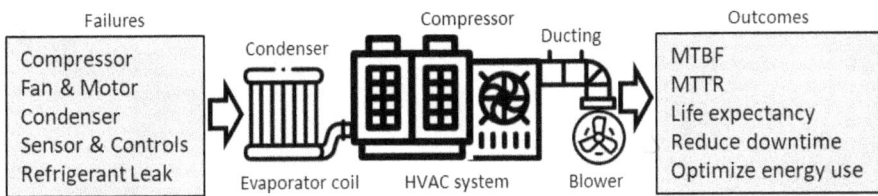

Figure 39: HVAC System PdM Example

Example: A simple example calculation is depicted below for a system like the one shown in Figure 39. Calculate the availability and failure rate for an HVAC system, we'll use the following formulas:

- Availability (A) = MTBF / (MTBF + MTTR)
- Failure Rate (λ) = 1 / MTBF

Given the following values:

- Life Expectancy (T) = 15 years = 15 * 365 * 24 hours
- MTTR = 8 hours
- MTBF = 4000 hours

Let's calculate the availability first:

Availability (A) = 4000 / (4000 + 8) = 4000 / 4008 $\approx$ 0.9980

The availability of the HVAC system is approximately 0.9980 or 99.80%.

Next, let's calculate the failure rate:

Failure Rate (λ) = 1 / 4000 = 0.00025

The failure rate of the HVAC system is 0.00025 failures per hour.

Please note that these calculations assume a constant failure rate over time and do not consider any other factors that may affect the availability and failure rate of the system.

About the Authors

Dr. G.E. Thompson specializes in remote data collection and analysis for predictive maintenance (PdM). He is the founding owner of a tracking vehicle location/current condition company and a renewable energy business. Greg is a retired Marine Officer and DoD engineering consultant. Presently, he is a senior electrical engineer for Andromeda Systems Incorporated. He holds advanced degrees in Engineering and Mathematics. A prolific writer and researcher, Greg has six U.S. patents.

Ron Wagner is the inventor of "Condition Based Logistics," the management system of logistics decision-making using conditions to trigger alternatives in support procedures based on conditions or events that occur in maintenance, distribution, transportation, and inventory systems. He is the current Chief Technology Officer (CTO) for Andromeda Systems Incorporated. Ron holds 6 U.S. patents in radio frequency identification, sensor networks, and telecommunications. A graduate of the United States Naval Academy, Class of 1974, Ron has a graduate degree in Material Management. He is a certified acquisition professional in program management, systems engineering, logistics, and industrial management.

Rob Ufford is the author of CERTIFY-IoT, a digital business transformation methodology emphasizing Lean operations enabled by real-time data. He was a key technology contributor to a global aerospace and defense company initiative that applied predictive maintenance technologies to military and commercial vehicles. After his years at Apple Computer, Rob became a serial entrepreneur, having co-founded four IoT wireless sensor and identification companies. A graduate of West Point, Rob holds advanced degrees in Computer Science, Artificial Intelligence, and Systems Management. He has five U.S. patents.

www.ingramcontent.com/pod-product-compliance
Lightning Source LLC
Chambersburg PA
CBHW070407200326
41518CB00011B/2102